This report contains the collective views of an international group of experts and does not necessarily represent the decisions or the stated policy of the United Nations Environment Programme, the International Labour Organisation, or the World Health Organization.

Environmental Health Criteria 111

TRIPHENYL PHOSPHATE

Published under the joint sponsorship of the United Nations Environment Programme, the International Labour Organisation, and the World Health Organization

First draft prepared by Dr A. Nakamura, National Institute for Hygienic Sciences, Japan

World Health Organization
Geneva, 1991

The **International Programme on Chemical Safety (IPCS)** is a joint venture of the United Nations Environment Programme, the International Labour Organisation, and the World Health Organization. The main objective of the IPCS is to carry out and disseminate evaluations of the effects of chemicals on human health and the quality of the environment. Supporting activities include the development of epidemiological, experimental laboratory, and risk-assessment methods that could produce internationally comparable results, and the development of manpower in the field of toxicology. Other activities carried out by the IPCS include the development of know-how for coping with chemical accidents, coordination of laboratory testing and epidemiological studies, and promotion of research on the mechanisms of the biological action of chemicals.

WHO Library Cataloguing in Publication Data

Triphenyl phosphate.

(Environmental health criteria ; 111)

1.Organophosphorus compounds - adverse effects 2.Organophosphorus compounds - toxicity I.Series

ISBN 92 4 157111 X (NLM Classification: QV 627)
ISSN 0250-863X

©World Health Organization 1991

Publications of the World Health Organization enjoy copyright protection in accordance with the provisions of Protocol 2 of the Universal Copyright Convention. For rights of reproduction or translation of WHO publications, in part or *in toto*, application should be made to the Office of Publications, World Health Organization, Geneva, Switzerland. The World Health Organization welcomes such applications.

The designations employed and the presentation of the material in this publication do not imply the expression of any opinion whatsoever on the part of the Secretariat of the World Health Organization concerning the legal status of any country, territory, city, or area or of its authorities, or concerning the delimitation of its frontiers or boundaries.

The mention of specific companies or of certain manufacturers' products does not imply that they are endorsed or recommended by the World Health Organization in preference to others of a similar nature that are not mentioned. Errors and omissions excepted, the names of proprietary products are distinguished by initial capital letters.

Printed in Finland
DHSS — Vammala — 5000

CONTENTS

ENVIRONMENTAL HEALTH CRITERIA FOR
TRIPHENYL PHOSPHATE

1. SUMMARY   11

   1.1 Identity, physical and chemical properties,
analytical methods   11
   1.2 Sources of human and environmental exposure   11
   1.3 Environmental transport, distribution, and
transformation   11
   1.4 Environmental levels and human exposure   12
   1.5 Effects on organisms in the environment   12
   1.6 Effects on experimental animals and *in vitro*
test systems   12
   1.7 Effects on humans   13

2. IDENTITY, PHYSICAL AND CHEMICAL PROPERTIES,
ANALYTICAL METHODS   14

   2.1 Identity   14
   2.2 Physical and chemical properties   15
   2.3 Conversion factor   16
   2.4 Analytical methods   16
      2.4.1 Sample extraction   16
      2.4.2 Clean-up procedures   20
      2.4.3 Gas chromatography and mass spectrometry   20
      2.4.4 Contamination of analytical reagents   20
      2.4.5 Other analytical methods   20

3. SOURCES OF HUMAN AND ENVIRONMENTAL
EXPOSURE   21

   3.1 Production levels and processes   21
   3.2 Uses   21

4. ENVIRONMENTAL TRANSPORT, DISTRIBUTION,
AND TRANSFORMATION   22

   4.1 Transport and transformation in the environment   22
      4.1.1 Release to the environment   22
      4.1.2 Fate in water and sediment   23

		4.1.3	Biodegradation	23
		4.1.4	Water treatment	24
	4.2	Bioaccumulation		24
		4.2.1	Fish	24
		4.2.2	Chironomid larvae	26
		4.2.3	Environmental fate in artificial pond water	28

5. ENVIRONMENTAL LEVELS AND HUMAN EXPOSURE ... 29

    5.1 Environmental levels ... 30
        5.1.1 Air ... 30
        5.1.2 Water ... 30
        5.1.3 Sediment and soil ... 34
        5.1.4 Fish ... 34
    5.2 General population exposure ... 35
        5.2.1 Food ... 35
        5.2.2 Drinking-water ... 35
        5.2.3 Human tissues ... 36
    5.3 Occupational exposure ... 36

6. EFFECTS ON ORGANISMS IN THE ENVIRONMENT ... 38

    6.1 Unicellular algae and fungi ... 38
    6.2 Aquatic organisms ... 38
    6.3 Insects ... 43

7. KINETICS AND METABOLISM ... 44

8. EFFECTS ON ANIMALS AND *IN VITRO* TEST SYSTEMS ... 45

    8.1 Single exposure ... 45
    8.2 Short-term exposure ... 45
    8.3 Skin irritation ... 47
    8.4 Reproduction ... 47
    8.5 Mutagenicity ... 47
    8.6 Carcinogenicity ... 48
    8.7 Neurotoxicity ... 48
    8.8 *In vitro* studies ... 49

9. EFFECTS ON HUMANS ... 51

10. EVALUATION OF HUMAN HEALTH RISKS AND
    EFFECTS ON THE ENVIRONMENT   52

    10.1 Evaluation of human health risks   52
         10.1.1 Exposure levels   52
         10.1.2 Toxic effects   52
    10.2 Evaluation of effects on the environment   53
         10.2.1 Exposure levels   53
         10.2.2 Toxic effects   53

11. RECOMMENDATIONS   55

    11.1 Recommendations for further research   55

REFERENCES   56

RESUME   65

EVALUATION DES RISQUES POUR LA SANTE
HUMAINE ET DES EFFETS SUR L'ENVIRONNEMENT   69

RECOMMANDATIONS   73

RESUMEN   74

EVALUACION DE LOS RIESGOS PARA LA SALUD
HUMANA Y DE LOS EFECTOS EN EL MEDIO AMBIENTE   77

RECOMENDACIONES   80

WHO TASK GROUP ON ENVIRONMENTAL HEALTH CRITERIA FOR TRIPHENYL PHOSPHATE

Members

Dr S. Dobson, Institute of Terrestrial Ecology, Monks Wood Experimental Station, Abbots Ripton, Huntingdon, Cambridgeshire, England *(Chairman)*

Dr S. Fairhurst, Medical Division, Health and Safety Executive, Bootle, Merseyside, England *(Joint Rapporteur)*

Ms N. Kanoh, Division of Information on Chemical Safety, National Institute of Hygienic Sciences, Setagaya-ku, Tokyo, Japan

Dr A. Nakamura, Division of Medical Devices, National Institute of Hygienic Sciences, Setagaya-ku, Tokyo, Japan

Dr M. Tasheva, Department of Toxicology, Institute of Hygiene and Occupational Health, Sofia, Bulgaria

Dr B. Veronesi, Neurotoxicology Division, US Environmental Protection Agency, Research Triangle Park, North Carolina, USA

Mr W.D. Wagner, Division of Standards Development and Technology Transfer, National Institute for Occupational Safety and Health, Cincinnati, Ohio, USA

Dr R. Wallentowicz, Exposure Assessment Application Branch, US Environmental Protection Agency, Washington, DC, USA *(Joint Rapporteur)*

Dr Shen-Zhi Zhang, Beijing Municipal Centre for Hygiene and Epidemic Control, Beijing, China

Observers

Dr M. Beth, Berufsgenossenschaft der Chemischen Industrie (BG Chemie), Heidelberg, Federal Republic of Germany

Dr R. Kleinstück, Bayer AG, Leverkusen, Federal Republic of Germany

Secretariat

Dr M. Gilbert, International Programme on Chemical Safety, Division of Environmental Health, World Health Organization, Switzerland (*Secretary*)

NOTE TO READERS OF THE CRITERIA DOCUMENTS

Every effort has been made to present information in the criteria documents as accurately as possible without unduly delaying their publication. In the interest of all users of the environmental health criteria documents, readers are kindly requested to communicate any errors that may have occurred to the Manager of the International Programme on Chemical Safety, World Health Organization, Geneva, Switzerland, in order that they may be included in corrigenda, which will appear in subsequent volumes.

* * *

A detailed data profile and a legal file can be obtained from the International Register of Potentially Toxic Chemicals, Palais des Nations, 1211 Geneva 10, Switzerland (Telephone No. 7988400 or 7985850).

ENVIRONMENTAL HEALTH CRITERIA FOR TRIPHENYL PHOSPHATE

A WHO Task Group meeting on Environmental Health Criteria for Triphenyl Phosphate was held at the British Industrial Biological Research Association (BIBRA), Carshalton, United Kingdom, from 9 to 13 October 1989. Dr S.D. Gangolli, Director, BIBRA, welcomed the participants on behalf of the host institution and Dr M. Gilbert opened the meeting on behalf of the three cooperating organizations of the IPCS (ILO, UNEP, WHO). The Task Group reviewed and revised the draft criteria document and made an evaluation of the risks for human health and the environment from exposure to triphenyl phosphate.

The first draft of this document was prepared by Dr A. Nakamura, National Institute for Hygienic Sciences, Japan. Dr M. Gilbert and Dr P.G. Jenkins, both members of the IPCS Central Unit, were responsible for the overall scientific content and editing, respectively.

ABBREVIATIONS

BCF	bioconcentration factor
EC	effective concentration
HPLC	high performance liquid chromatography
LC	lethal concentration
LD	lethal dose
ND	not detected
OPIDN	organophosphate-induced delayed neuropathy
TAP	triaryl phosphate
TCP	tricresyl phosphate
TLC	thin-layer chromatography
TPP	triphenyl phosphate

1. SUMMARY

1.1 Identity, physical and chemical properties, analytical methods

Triphenyl phosphate (TPP) is a non-flammable, non-explosive, colourless, crystalline substance. Its partition coefficient between octanol and water (log P_{ow}) is 4.61-4.76. At normal ambient temperatures, it hydrolyses rapidly in alkaline solution, producing diphenyl phosphate and phenol, but very slowly in acidic or neutral solutions.

The analytical method of choice is gas-liquid chromatography with nitrogen-phosphorus sensitive or flame photometric detection. The detection limit in water is about 20 ng/litre.

1.2 Sources of human and environmental exposure

TPP is manufactured from phosphorus oxychloride and phenol. It is used as a flame retardant in phenolic and phenylene-oxide-based resins for the manufacture of electrical and automobile components and as a non-flammable plasticizer in cellulose acetate for photographic films. It is also a component of hydraulic fluids or lubricant oils and has a number of other minor uses.

Exposure of the general population through normal use can be regarded as minimal.

1.3 Environmental transport, distribution, and transformation

Triaryl phosphates enter into the aquatic environment mainly via hydraulic fluid leakage as well as by leaching from plastics and, to a minor extent, from manufacturing processes. Because of its low water solubility and relatively high soil adsorption coefficient, TPP is rapidly adsorbed on river (or pond) sediments. Its biodegradation in aqueous environments is rapid.

The degradation of TPP involves a stepwise enzymatic hydrolysis to orthophosphate and phenolic moieties.

Summary

The bioconcentration factors (BCF) measured for several species of fish range from 6 to 18 900 and the depuration half-life ranges from 1.2 to 49.6 h.

The release of TPP from production sites to the air represents a source of human exposure in the occupational environment. The combustion of plastics and volatilization from plastics or water surfaces may also be a major pathway to the atmosphere.

1.4 Environmental levels and human exposure

TPP has been widely found in air, water, sediment, and aquatic organisms, but levels in environmental samples are low. The maximum levels reported are 23.2 ng/m^3 in air, 7900 ng/litre in river water, 4000 ng/g in sediment, and 600 ng/g in fish.

1.5 Effects on organisms in the environment

The growth of algae is completely inhibited at TPP concentrations of 1 mg/litre or more but is stimulated at lower concentrations (0.1 and 0.05 mg/litre). The nitrogenase activity of *Anabaena flos-aquae* decreases in a dose-dependent manner from 84% at 0.1 mg/litre to 68% at 5.0 mg/litre.

TPP is the most acutely toxic of the various triaryl phosphates to fish, shrimps, and daphnids. The acute toxicity index of TPP for fish (96-h LC_{50}) ranges from 290 mg/litre in bluegills to 0.36 mg/litre in rainbow trout. The large difference in EC_0 values between trout and fathead minnows may be due to the difference in their ability to metabolize TPP. Sublethal effects on fish include morphological abnormalities such as congestion, degeneration, and haemorrhage from the smaller blood vessels (mainly in the gills) and behavioural abnormalities. The immobility of fish exposed to 0.21-0.29 mg per litre completely disappeared within 7 days when the fish were transferred to clean water.

1.6 Effects on experimental animals and *in vitro* test systems

The oral LD_{50} of TPP has been estimated to be >6.4 g/kg in rats and >2.0 g/kg in chickens.

TPP doses ranging from 0.5 to 2 g/kg were well tolerated by rabbits after intramuscular injection and by chickens after oral administration. In a 35-day feeding study, depression of body weight gain and increase in liver weight were observed at a dose of TPP in male Holtzman rats.

TPP was not teratogenic in Sprague-Dawley rats at dose levels up to 690 mg/kg body weight. No reproduction studies have been reported.

There are no data on the mutagenicity of TPP from well-validated tests, and there has been no adequate carcinogenicity study.

TPP did not cause delayed neurotoxicity following single subcutaneous exposures in cats (up to 1 g/kg) or in a 4-month study in Sprague-Dawley rats at dose levels up to 1% in the feed.

No immunotoxic effects were reported from a 120-day study in rats fed dose levels up to 1% in the feed.

1.7 Effects on humans

While a statistically significant reduction in red blood cell cholinesterase has been reported in some workers, there has been no evidence of neurological disease in workers in a TPP-manufacturing plant. There have been no reports of delayed neurotoxicity in cases of TPP poisoning. Contact dermatitis due to TPP has been described.

2. IDENTITY, PHYSICAL AND CHEMICAL PROPERTIES, ANALYTICAL METHODS

2.1 Identity

Chemical structure:

[Structure of triphenyl phosphate: three phenyl groups connected via O to a central P=O]

Molecular formula:	$C_{18}H_{15}O_4P$
Relative molecular mass:	326.3
CAS chemical name:	Phosphoric acid, triphenyl ester
CAS registry number:	115-86-6
RTECS registry number:	TC8400000
Synonyms:	Triphenyl phosphate; Triphenylphosphate; TPP
Trade name:	Phosflex TPP®; Disflamoll TP®; Celluflex TPP®

Manufacturers and suppliers (Modern Plastics Encyclopedia, 1975):

Ashland Chemical Co.; Celanese Co.; Daihachi Chemical Industry Co., Ltd.; East Coast Chemicals Co.; B.F. Goodrich Chemical Co.; Mobay Chemical Co.; Monsanto

Chemical Co.; Rhone-Poulenc Co.; Showa Ether Co., Ltd.; Stauffer Chemical Co.

2.2 Physical and chemical properties

The physical properties of TPP are listed in Table 1.

Table 1. Physical properties of TPP

Physical state	crystalline solid
Colour	colourless
Odour	very slightly aromatic
Melting point (°C)	49-50[a]; 49[b]; 49.2[c]
Boiling point (°C)	245 (11 mmHg)[a,b]; 220 (5 mmHg)[c]; 234 (5 mmHg)[d]; 370[e]
Relative density	1.185-1.202 (25 °C)[c]; 1.185 (25 °C)[d]; 1.2055[b]
Refractive index (at 25 °C)	1.552-1.563[c]
Flash point (°C)	220[b]; 225[c]
Viscosity (cSt)	11 (50 °C); 9.9 (55 °C)[c]
Vapour pressure (mmHg)	0.15 (150 °C); 1.90 (200 °C)[c]; 1.0 (193.5 °C)[e]
Henry's Law constant	1.8-3.6 × 10^{-7} atm-m^3/mol
Solubility in organic solvents	soluble in benzene, chloroform, ether, acetone; moderately soluble in ethanol[a]
Solubility in water (mg/litre)	1.9[f]; 0.73[g]; 2.1 (±0.1)[h]
Octanol-water partition coefficient (log P_{ow})	4.63[f]; 4.61[i]; 4.76[j]

[a] Windholz (1983)
[b] Hine et al. (1981)
[c] Modern Plastics Encyclopedia (1975)
[d] Lefaux (1972)
[e] Sutton et al. (1960)
[f] Saeger et al. (1979)
[g] Hollifield (1979)
[h] Ofstad & Sletten (1985)
[i] Kenmotsu et al. (1980b)
[j] Sasaki et al. (1981)

TPP is non-flammable and non-explosive. It begins to decompose at about 600 °C but is not completely degraded even at 1000 °C in inert gas. Under these conditions, TPP yields aromatic hydrocarbons (naphthalene, biphenyl, phenanthrene, anthracene, etc.), oxygenated aromatic compounds (phenol, dibenzofuran, diphenyl ether) and phosphoric oxides (ortho-, pyro-, meta-, and poly-phosphoric acids). With a large excess of air, complete combustion to carbon dioxide is accomplished within the temperature range 800-900 °C (Lhomme et al., 1984).

At ordinary temperature, TPP is hydrolysed very slowly in acidic and neutral solutions but rapidly in alkaline solutions. The hydrolysis rate constants and half-lives are summarized in Table 2. In studies by Barnard et al. (1961), alkaline hydrolysis of TPP yielded diphenyl phosphate, but further hydrolysis to monophenyl phosphate and phosphoric acid was not observed under the experimental conditions used. Under strong acidic conditions and at high temperature (100 °C), TPP readily hydrolyses to give phosphoric acid (Barnard et al., 1966).

In studies by Finnegan & Matson (1972), the photolysis of TPP yielded biphenyl (2%), the recovered ester amounting to 48%. The quantum yield for biphenyl formation was 6×10^{-4}.

2.3 Conversion factor

Triphenyl phosphate   1 ppm = 13.35 mg/m³ air

2.4 Analytical methods

Analytical methods for determining TPP in air, water, sediment, fish, and biological tissues are summarized in Table 3. General procedures for TPP analysis are similar to those for tricresyl phosphate (TCP) (WHO, 1990). The detection limit of TPP in water is approximately 20 ng/litre.

Diphenyl phosphate, a hydrolysis product of TPP, has been determined in sediment by extraction with aqueous methanol, clean-up with XAD-2 resin and C-18 bonded silica cartridge, butylation, and gas chromatographic determination (Muir et al., 1983b).

TPP is present in several commercial triaryl phosphates, e.g., Santicizer-140®, Pydraul 50E® (Monsanto Co.), Fyrquel GT®, and Phosflex 41-P® (Stauffer Chemical Co.) (Deo & Howard, 1978). When TPP is identified, other triaryl phosphates are often detected at the same time.

2.4.1 Sample extraction

TPP is extracted from water, sediment, fish, and air along with TCP. WHO (1990) gives details of methods.

Table 2. Hydrolysis rate constants and half-lives of TPP in aqueous solution

Solution	Temperature (°C)	pH	K_1 first order (sec^{-1})	Rate constants K_2 second order ($M^{-1}.sec^{-1}$)	K_1' pseudo-first order	Half-life	Reference
Water	27	alkaline		2.7×10^{-1}			Wolfe (1980)
60% dioxane-water	0	alkaline		2.35×10^{-3}			Barnard et al. (1961)
	10.1	alkaline		4.77×10^{-3}			Barnard et al. (1961)
	24.7	alkaline		1.06×10^{-2}			Barnard et al. (1961)
	35	alkaline		2.32×10^{-2}			Barnard et al. (1961)
NaOH (0.1 mol/litre)/acetone (1:1)	22	13.0				0.49 h	Muir et al. (1983a)
H_3BO_4/NaOH buffer	25	9				3 days	Mayer et al. (1981)
Buffered water	21 ± 2	8.2			9.3×10^{-2}/day	7.5 days	Howard & Deo (1979)
	9.5					1.3 days	Howard & Deo (1979)
Dioxane-water (3:1)	100	neutral				130 days	Barnard et al. (1961)
KH_2PO_4/Na_2HPO_4	25	7			6.0×10^{-8}/day	19 days	Mayer et al. (1981)
$KHC_8H_4O_4$/NaOH buffer	25	5				28 days	Mayer et al. (1981)
Dioxane-water (3:2)	100	neutral	6×10^{-8}				Barnard et al. (1966)
		0.122M $HClO_4$	1.43×10^{-5}				Barnard et al. (1966)
		1.21M $HClO_4$	10.8×10^{-5}				Barnard et al. (1966)
		3.02M $HClO_4$	6.45×10^{-5}				Barnard et al. (1966)

Table 3. Methods for the determination of TPP

Sample type	Sampling method extraction/clean-up	Analytical method	Limit of detection	Applicability	Reference
Workplace air	collect with Millipore filter, extract with ethanol	GC/FPD	1 µg per sample	TCP & TPP	US NIOSH (1982)
Environment air	trap with glycerol-Florisil column, eluate with methanol, add water, and extact with hexane	GC/FPD	1 ng/m^3	simultaneous method for trialkyl/aryl phosphates	Yasuda (1980)
Air	collect by aspiration through ethanol, hydrolyse with NaOH; the resultant phenols are reacted with p-$O_2NC_6H_4N_2^+$ and separated with silica gel plate	TLC	5 ng/plate	TCP & TPP	Druyan (1975)
Drinking-water	adsorb with XAD-2 resin, eluate with acetone-hexane or acetone	GC/NPD GC/MS	1 ng/litre	method for low level trialkyl/aryl phosphates	Lebel et al. (1979, 1981)
River or sea water	extract with methylene chloride or benzene	GC/NPD GC/FPD GC/MS	0.02 µg/litre (TPP) 0.05 µg/litre (TCP)	simultaneous method for trialkyl/aryl phosphates	Kenmotsu et al. (1980a, 1981b, 1982b) Muir et al. (1981) Ishikawa et al. (1985)

Table 3 (contd).

Sample	Method	Detection	Limit	Notes	Reference
Farm pond sediment	reflux with methanol-water (9+1) or methylene chloride-methanol (1+1), clean-up by acid alumina column chromatography	GC/NPD	1 ng/g	simultaneous method for triaryl phosphates	Muir et al. (1980, 1981)
River or sea sediment	extract with acetonitrile or acetone, clean-up by charcoal or Florisil column chromatography	GC/FPD GC/MS	5 ng/g	simultaneous method for trialkyl/aryl phosphates	Kenmotsu et al. (1980a, 1981b, 1982a, 1982b, 1983) Ishikawa et al. (1985)
Fish	extract with hexane or methanol, clean-up by gel permeation column chromatography and acid alumina column chromatography	GC/NPD GC/MS	1 ng/g	simultaneous method for triaryl phosphates	Muir et al. (1980, 1981, 1983)
Fish	extract with acetonitrile and methylene chloride, clean-up by acetonitrile-hexane partitioning, charcoal column chromatography, concentrated sulfuric acid extraction and Florisil column chromatography	GC/FPD GC/MS	5 ng/g	simultaneous method for trialkyl/aryl phosphates	Kenmotsu et al. (1980a)
Human adipose tissues	extract with benzene or acetone-hexane (15 + 85), clean-up by gel permeation chromatography and Florisil column chromatography	GC/NPD GC/FPD GC/MS	1 ng/g	simultaneous method for trialkyl/aryl phosphates	Lebel & Williams (1983)

2.4.2 Clean-up procedures

Clean-up procedures for TPP are similar to those for TCP (WHO, 1990). It is difficult to separate TPP from other triaryl phosphates by Florisil or gel permeation chromatography.

2.4.3 Gas chromatography and mass spectrometry

TPP is analysed simultaneously with TCP. GC and mass spectrometry procedures are described in WHO (1990).

2.4.4 Contamination of analytical reagents

Triaryl phosphates (TAPs) are widely used as flame retardants in plastics and hydraulic fluids. Their widespread use and release into the environment produces trace contamination of reagents used for analysis. Trace amounts of TPP have been found in Super Q water (Williams & Lebel, 1981), Corning water (Lebel et al., 1981), hexane, acetonitrile, and methylene chloride (Daft, 1982). TAPs have also been found in cyclohexane (Bowers et al., 1981), hexane (Hudec et al., 1981), and analytical grade filters (Daft, 1982). Care must be taken to avoid contamination of reagents in order to obtain reliable data in trace analysis of TPP.

2.4.5 Other analytical methods

A colorimetric method has been developed for determination of TPP in air (Druyan, 1975), but the interference by other TAPs was not investigated. Thin-layer chromatography (TLC) has been used for the determination of TPP in air (Druyan, 1975) and in plastics (Peereboom, 1960; Braun, 1965). The octanol/water partition coefficient of TPP has been determined by reversed phase TLC (Renberg et al., 1980). It is difficult to separate the various TAPs by TLC (Bloom, 1973). Tittarelli & Mascherpa (1981) described a highly specific HPLC detector for TAPs using a graphite furnace atomic absorption spectrometer. In general, TLC and HPLC have not been used as widely as GLC for the analysis of TPP.

3. SOURCES OF HUMAN AND ENVIRONMENTAL EXPOSURE

3.1 Production levels and processes

TPP does not occur naturally in the environment. Figures concerning total world production are not available, but 7250 tonnes was produced in the USA in 1977 (Boethling & Cooper, 1985) and 3750 tonnes in Japan in 1984.

TPP is produced from phosphorus oxychloride and phenol.

3.2 Uses

TPP was used, in Japan in 1984, as a flame-retardant in phenolics and phenylene-oxide-based resin for the manufacture of electrical and automobile components (3200 tonnes), as a non-flammable plasticizer in cellulose acetate for photographic films (500 tonnes), and for other miscellaneous purposes (50 tonnes)[a]. Other uses of TPP are as a non-combustible substitute for camphor in celluloid (which renders acetylcellulose, nitrocellulose, airplane "dope", etc. stable and fireproof), for impregnating roofing paper, and as a plasticizer in lacquers and varnishes (Windholz, 1983). It is also used as a plasticizer in vinyl automotive upholstery (Ahrens et al., 1978) and in cellulose acetate articles (Pegum, 1966).

TPP is also found as a component of hydraulic fluids and lubricant oils (WHO, 1990; Table 6), and of other triaryl phosphate esters: methyl diphenyl phosphate (triphenyl phosphate content, ≈ 5%); 2-ethylhexyl diphenyl phosphate (≈ 5%); trixylenyl phosphate (≈ 5%); isodecyl diphenyl phosphate (≈ 45%); cresyl diphenyl phosphate (≈ 45%); isopropylphenyl diphenyl phosphate (≈ 45%) (Daft, 1982).

---

[a] Personal communication to IPCS from the Association of the Plasticizer Industry of Japan (1985).

4. ENVIRONMENTAL TRANSPORT, DISTRIBUTION, AND TRANSFORMATION

Summary

TPP has been found in various environmental media, but usually at low levels. It may be released by leakage at sites of production and use and by the combustion of plastics. No figures are available on the amounts released into the environment.

The solubility of TPP in water is low, and it is readily adsorbed onto sediment.

The rate of biodegradation in water is dependent on water quality (1 mg/litre was degraded in 4 days in River Mississippi water). Little or no degradation occurs in heat-sterilized river water. The degradation pathway is reported to involve stepwise enzymatic hydrolysis.

Water treatment techniques, both for waste water and drinking-water, reduce TPP levels by at least an order of magnitude.

Bioaccumulation data are available from laboratory studies, but should be considered to represent a bioaccumulation potential. Depuration, as measured by clearance rate constant, is higher for rainbow trout than for fathead minnows by about 50%.

4.1 Transport and transformation in the environment

4.1.1 Release to the environment

The release of TPP into the air at production sites represents a potential source of human contamination. It has been suggested that, since TPP in the reactor and purification vessels is hot, mist and vapour coming from leaks in the reactor and from the open receiving tank are the main source of TPP in the air (Sutton et al., 1960) (see also section 5.2). Recent figures however are not available. A low concentration (0.057 mg/m^3) of TPP was detected near a zinc die cast machine where hydraulic fluids were used (US NIOSH, 1980).

Combustion of plastics or volatilization from plastics or water surface may also be a major pathway to the atmosphere. Vick et al. (1978) found TPP emitted in the vapour phase and on particulate matter from a utility plant. The concentrations were not reported.

The entry of TPP into the aquatic environment is thought to occur principally via hydraulic fluid leakage, as well as by leaching from vinyl plastics and, to a minor extent, from manufacturing processes (Ahrens et al., 1978; Mayer et al., 1981; WHO, 1990).

4.1.2 Fate in water and sediment

The solubility of TPP in water is low (Table 1).

Monitoring studies have shown trialkyl and triaryl phosphates to be present in water and sediment sampled near major industrialized sites (Konasewich et al., 1978; Sheldon & Hites, 1978, 1979; Mayer et al., 1981; Williams & Lebel, 1981; Aldous, 1982; Williams et al., 1982; Ishikawa et al., 1985). The adsorption coefficient of TPP on marine sediment was found to be 59 (Kenmotsu et al., 1980b). Muir et al. (1982) showed rapid equilibrium of TPP with the bottom sediment in a shallow pond (depth 0.5 m) within 10 h.

4.1.3 Biodegradation

TPP (200 μg) was completely degraded within 4 days in 200 ml of River Mississippi (USA) water at room temperature (Saeger et al., 1979). Howard & Deo (1979) measured the degradation rate constants for TPP in non-sterilized natural water (Seneca River and Lake Ontario, USA). Little degradation occurred for the first two days, followed by a loss more rapid than in distilled water at comparable pH. After two days, the pseudo-first-order rate constants at pH 8.2 were 0.64 and 0.34 days^{-1} for the two natural water samples, and 0.093 days^{-1} for distilled water. The rapid degradation (99.2% in 7 days) of TPP (1 mg/litre) was also found in a river die-away study using Neya and Oh River water (Osaka, Japan), whereas no degradation was observed during 15 days in heat-sterilized river water (Hattori et al., 1981). In clear non-sterilized sea water, the degradation was very slow (35.1% in 14 days) (Hattori et al., 1981).

Primary biodegradation rates from semicontinuous activated sludge studies generally show the same trend in degradation rates as river die-away studies; TPP (3-13 mg per litre, 24-h feed) revealed 96% (± 2%) degradation (Saeger et al., 1979). The ultimate biodegradability was measured using the apparatus and procedure developed by Thompson & Duthie and modified by Sturm; the theoretical carbon dioxide evolution from TPP (18.3 mg/litre) was 81.8% (Saeger et al., 1979).

The degradation pathway for TPP is reported to involve stepwise enzymatic hydrolysis to orthophosphate and phenolic moieties. The phenol would be expected to undergo further degradation (Barrett et al., 1969; Pickard et al., 1975).

4.1.4 Water treatment

Data from FMC Corporation (USA) show that TPP (0.74 mg per litre) in waste water was reduced to 0.07 mg/litre in the effluent water by biological treatment (Boethling & Cooper, 1985). TPP was reduced from 16 µg/litre to 2 µg/litre by classical secondary treatment methods, and from 0.2 µg/litre to 0.03 µg/litre by standard techniques for drinking-water treatment (Sheldon & Hites, 1979). Fukushima & Kawai (1986) also reported that TPP (0.054-2.12 µg/litre) in untreated water was reduced to 0.005-0.082 µg/litre by conventional waste water treatment.

4.2 Bioaccumulation

4.2.1 Fish

Data on the bioconcentration and depuration of TPP are given in Table 4. None of the exposures were considered to be representative of realistic environmental levels. Moreover the bioconcentration factor (BCF) measured in the laboratory must be considered to represent a bioaccumulation potential rather than an absolute bioaccumulation factor (Veith et al., 1979).

Several equations have been presented to predict the bioconcentration factors of organic chemicals in various fish strains using octanol-water partition coefficient

Table 4. Bioaccumulation and clearance of TPP by fish

Species	Temp (°C)	Flow/ stat	Analytical method[a]	BCF (k_1/k_2)	Exposure concentration (mg/litre)	Uptake rate (k_1, h^{-1})	Clearance rate ($k_2 \times 10^3$, h^{-1})	Depuration half life (h)	Reference
Killifish (*Oryzeas latipes*)	25	Stat	GC-FPD	157-390	1			1.2	Sasaki et al. (1982)
		Stat	GC-FPD	250-500	0.25				Sasaki et al. (1981)
		Flow	GC-FPD	84-193	1				Sasaki et al. (1982)
Rainbow trout (*Salmo gairdneri*)	10	Stat	TR	1368 ± 329[b]	0.005-0.05	9.7[c]	11.6-17.4	42.5	Muir et al. (1983a)
			TR	573 ± 97[c]			17.7[d]		
			TR	931 ± 122[d]			20.7		
			HER	324 ± 99[c]					
	10	Stat	TR	2590[b]	0.05	43.36	17.9 (fast)		Muir et al. (1980b)
				18 900[b]			2.45 (slow)		
	12	Stat	GC-FPD	271			12.96		Sitthichaikasem (1978)
Fathead minnows (*Pimephales promelas*)	10	Stat	TR	1743 ± 282[b]	0.005-0.05	15.4[d]	7.6-14.0	49.6	Muir et al. (1983a)
			TR	561 ± 115[c]			7.2[c]		
			TR	218 ± 55[d]					
			HER	420 ± 25[c]					
Goldfish (*Carassius auratus*)	25	Stat	GC-FPD	6-11				30.0	Sasaki et al. (1981)

[a] GC-FPD = gas chromatography (flame photometric detector) after suitable extraction; TR = total radioactivity; HER = hexane-extractable radioactivity.
[b] BCF was calculated by the "initial rate method".
[c] The static test method was used (Zitko, 1980).
[d] K_1 and k_2 were derived by non-linear regression calculation.

(P_{ow}) or water solubility (Neely et al., 1974; Lu & Metcalf, 1975; Kanazawa, 1978; Veith et al., 1979; Sasaki et al., 1982).

Of six tissues of fish exposed with ^{14}C-triphenyl phosphate, liver had the highest concentration (10 µg/g at 4 h post-treatment) and also showed the highest rate of ^{14}C-triphenyl phosphate depuration. The rapid clearance from liver suggests extensive TPP metabolism. Rates of uptake of radioactivity (µg/g tissue per h) by the six tissues were as follows: liver, 2.75; kidney, 2.01; caeca, 0.62; intestine, 0.53; muscle, 0.45; blood, 0.56 (Muir et al., 1980b).

Clearance of TPP was biphasic with more rapid rates of clearance in the first 6 days after transfer to clean water, especially in the case of rainbow trout (Muir et al., 1983a). The clearance rate constant was higher for rainbow trout than for fathead minnows by about 50% (Muir et al., 1983a).

4.2.2 Chironomid larvae

Muir et al. (1983b) studied the accumulation of TPP by *Chironomus tentans* larvae exposed to water and sediment spiked with ^{14}C-triphenyl phosphate. The overall accumulation has been described by the following equation:

$$dC_c/dt = k_1(C_w) - k_2(C_c) + k_3(C_s)$$

where C_c is the concentration in the larvae, C_w the concentration in water, C_s the concentration in sediment, k_1, k_3 are uptake rate constants, and k_2 is the elimination rate constant.

The rate constant k_3 has been described by the following equation:

$$k_3 = k_1(C_w/C_s) \times (CF_s - CF_w)/(CF_w)$$

where CF_s is the equilibrium concentration factor for larvae in sediment and CF_w is the equilibrium concentration factor for larvae in water.

The results are summarized in Table 5.

The relative contribution of sediment and water to the body burdens observed in larvae (24-h exposure) was esti-

Table 5. Uptake rate constants (k_1) calculated by use of a first-order kinetic model and concentration factors for uptake of TPP by Chironomid larvae

System	High concentration (500 µg/kg sediment)			Low concentration (50 µg/kg sediment)		
	k_1 (h^{-1})	CF_w	CF_s	k_1 (h^{-1})	CF_w	CF_s
Pond sediment	0.4 ± 0.1	6 ± 0	12 ± 10	0.6 ± 0.2	12 ± 4	18 ± 8
River sediment	2.1 ± 0.8	45 ± 17[a]	78 ± 34[a]	4.2 ± 1.6	88 ± 34	90 ± 41
Sand sediment	3.3 ± 1.0	64 ± 33[b]	173 ± 69[b]	10.4 ± 3.1	208 ± 62[b]	138 ± 17[b]

[a] Significant difference (P = 0.05) between mean larval concentrations in water and in sediment using the t-test.
[b] Significant difference (P = 0.01) between mean larval concentrations in water and in sediment using the t-test.

mated by calculating uptake rates for water $(k_1)(C_w)$ and for sediment $(k_3)(C_s)$. The results indicate that contributions from water and sediment were roughly equivalent for most sediments for a sediment-to-water ratio of 1:5. The authors noted however:

"A greater ratio of water to sediment would tend to increase the contribution of uptake from water proportionally. A water-to-sediment ratio of 100:1 would reduce the contribution of sediment uptake to 10%, and would make it difficult to show significant differences between water and sediment experiments".

Initial elimination rate constants and half-lives of TPP for larvae exposed to different sediment-water systems are shown in Table 6.

Table 6. Elimination rate constants and half-lives of TPP for Chironamid larvae

System	Elimination rate k_2 (h^{-1})	Half-life (h)
Pond sediment	0.023 ± 0.012	30.4 ± 16.1
River sediment	0.011 ± 0.004	62.7 ± 24.5
Sand sediment	0.016 ± 0.010	44.4 ± 28.0
River water	0.039 ± 0.013	17.6 ± 6.0

4.2.3 Environmental fate in artificial pond water

The environmental fate of radiolabelled TPP (60 µg per litre) in artificial pond water was studied by Muir et al. (1982). The radioactivity observed in each pond compartment is shown in Table 7. Small losses of TPP by volatilization were thought to occur, but this was not confirmed by direct measurement above the water surface. Despite differences in fish species and water temperatures between laboratory and field experiments, the observed body burdens of TPP were similar, for the first 24 h of the experiment, to those predicted on the basis of laboratory data. However, at 72 and 240 h, the predicted values were higher than those observed. These results were considered to reflect more rapid clearance of TPP by fathead minnows than by rainbow trout.

Table 7. Percentage of TPP radioactivity in the various compartments of a pond

Time(h)	Water	Sediment	Duckweed	Fish	Total
10	74	29	1.4	2.7	107.1
24	52	34	1.4	3.4	90.8
32	31	-	-	-	-
48	34	43	1.2	1.3	79.5
72	28	33	0.9	0.9	62.8
120	23	40	0.5	0.6	64.1
240	13	36	0.5	0.5	50.0

5. ENVIRONMENTAL LEVELS AND HUMAN EXPOSURE

Summary

TPP has been widely found in air, water, sediment, and aquatic organisms, but only at low levels. Higher levels have been found only in sediments near industrialized areas.

Ambient air concentrations of TPP in rural areas range from 0.5 to 1.4 ng/m^3 and in urban areas from 0.9 to 14.1 ng/m^3.

TPP levels in surface water from 3 to 700 ng/litre have been measured, values up to 7900 ng/litre occurring near facilities producing either aryl phosphates or hydraulic fluids containing TPP. These high values probably result from TPP bound to suspended sediment. Reported drinking-water levels are several orders of magnitude lower (0.3-30 ng/litre), suggesting that TPP is removed by adsorption to filtration media in water-treatment plants. No TPP has been detected in potable water from wells. There is no information available on levels in groundwater.

Levels of TPP in river and marine sediment range from 0.2 to 200 ng/g, but values of up to 4000 ng/g have been reported in sediments near manufacturing sites for automobile parts. There has been one report of TPP being detected in agricultural soils, but no values were given.

In fish and shellfish, TPP tissue levels from 2 to 150 ng per g have been reported. No TPP has been detected in human adipose tissue and there are no data on TPP for any other species.

Exposure to humans can occur by several routes, including the ingestion of contaminated drinking-water, fish, shellfish, and other foodstuffs. US FDA total-diet studies have found average daily intake levels of 0.3-4.4, 1.2-1.6, and 0.5-1.6 ng per kg body weight per day for infants, toddlers, and adults, respectively. It should be noted that TPP occurred in less than 1% of the foods in these diets.

Occupational exposure can occur in manufacturing industries and other areas such as automobile or aircraft facilities handling hydraulic fluids. Levels of 0.008 to 29.6 mg/m^3 have been detected in air, the highest values occurring at TPP-manufacturing sites.

5.1 Environmental levels

TPP has been found widely in air, water, sediment, and aquatic organisms. The levels of TPP in environmental samples are low (Table 8), although moderately high levels have often been found in sediment collected near heavily industrialized areas (Table 9 and 10).

5.1.1 Air

Yasuda (1980) measured the distribution of various organic phosphorus compounds in the atmosphere above the eastern Seto Inland Sea, Japan, and found TPP at levels of 0.5-1.4 ng/m^3 in 3 out of 4 samples. He also measured concentrations of phosphate esters in the atmosphere above the Dogo Plain and Ozu Basin of Western Shikoku, Japan, these being mainly agricultural areas. TPP was detected only in the urban air of a middle-size city (Matsuyama) at levels of 0.9-14.1 ng/m^3.

5.1.2 Water

Although there have been many studies of TAPs in water, TPP has not often been detected in natural waters. According to the annual reports of the Environment Agency of Japan, TPP has not been detected in river or sea water at any sampling points in Japan. Detection limits varied from 20 to 200 ng/litre at the various laboratories (EAJ, 1977, 1981). Kawai et al. (1978) detected TPP in river water sampled in Osaka, Japan, at levels of 50-700 ng per litre, and Ishikawa et al. (1985) detected levels of 13-31 ng/litre in 5 out of 16 samples of river water in Kitakyushu City, Japan, but none in sea water. Both cities are located in the most heavily industrialized area of Japan. In Tokyo, Japan, TPP was not found in river or sea water (detection limit: 20 ng/litre) by Wakabayashi (1980), whereas a level of 3 ng/litre was detected in sea water by Sugiyama & Tanaka (1982).

High concentrations of TAPs have frequently been detected in river water sampled near producer and user sites. Sheldon & Hites (1978) found 100-300 ng TPP/litre in 2 out of 5 samples of Delaware River (USA) water

Table 8. Concentrations of TPP in environmental air, water, sediment, and fish at various locations

Year	Location	Sample	Concentration[a]	Number of samples (detected/analysed)	Reference
1975	Japan (various locations)	river and sea water	ND (20-200 ng/litre)	(0/100)	EAJ (1977)
		river and sea sediment	ND (2-50 ng/g)	(0/100)	
		fish	ND (5-50 ng/g)	(0/100)	
1976	Osaka (Japan)	river water	50-700 ng/litre	(11/13)	Kawai et al. (1978)
1976	Shikoku (Japan)	atmosphere	0.9-14.1 ng/m^3	(4/19)	Yasuda (1980)
1977	Eastern Seto Inland Sea (Japan)	atmosphere	0.5-1.4 ng/m^3	(3/4)	
1978	Eastern Ontario water treatment plant (Canada)	drinking-water	0.3-2.6 ng/litre	(12/12)	Lebel et al. (1981)
1978	Tokyo (Japan)	river water	ND (20 ng/litre)	(0/12)	Wakabayashi (1980)
		sea water	ND (20 ng/litre)	(0/3)	
		river sediment	0.7-3.3 ng/g	(10/15)	
		sea sediment	0.2-0.3 ng/g	(2/3)	
1979	Canada (various locations)	drinking-water	0.3-8.6 ng/litre	(20/60)	Williams & Lebel (1981)
1980	Great Lake (Canada)	drinking-water	0.2-4.8 ng/litre	(11/12)	Williams et al. (1982)
1980	Kitakyushu City (Japan)	river water	13-31 ng/litre	(3/16)	Ishikawa et al. (1985)
		sea water	ND (10 ng/litre)	(0/9)	
		sea sediment	ND (5 ng/g)	(0/6)	
1980	Seto Inland Sea (Japan)	fish and shellfish	2-6 ng/g	(12/41)	Kenmotsu et al. (1981a)
1981	Tokyo Bay (Japan)	sea water	3 ng/litre		Sugiyama & Tanaka (1982)
NR	USA	drinking-water	10-120 ng/litre		Muir (1984)

[a] Figures in parentheses are detection limits
ND = not detected; NR = not reported

Table 9. Concentration of TPP in water, sediment, and fish muscle at industrialized and non-industrialized sites in the USA[a]

Location	Water (ng/litre)	Sediment (ng/g)	Fish (ng/g)
Waukegan Harbor, Illinois	ND (0/5)	10 (2/3)	ND (0/13)
Waukegan Bay, Illinois	ND (0/4)	ND (0/3)	NR
Upper Saginaw River, Michigan	100 (3/3)	10 (3/3)	ND (0/12)
Saginaw River at Lake Huron	600-700 (4/4)	1000-4000 (3/3)	100 (1/10)
Illinois River, Grafton Illinois	100 (4/4)	ND (0/3)	ND (0/4)
Missouri River, Fenton, Missouri	ND (0/4)	ND (0/3)	ND (0/9)
Missouri River at Chesterfield, Missouri	100 (1/4)	ND (0/4)	NR
Missouri River, Halls Ferry, Missouri	100-200 (3/3)	ND (0/3)	NR
Mississippi River above St. Louise, Missouri	ND (0/5)	NR	500 (1/3)
Mississippi River at St. Louise, Missouri	100-7900 (9/15)	100 (2/6)	NR
Mississippi River below St. Louise, Missouri	100-400 (3/3)	ND (0/3)	100 (1/4)
Kanawha River, Winfield, W. Virginia	100-800 (3/3)	20-200 (3/6)	100-600 (13/27)
San Francisco Bay, California	100 (2/5)	NR	NR

[a] From: Mayer et al. (1981); measurements were made from November 1977 to May 1978; figures in parentheses indicate number of samples (detected/analysed); detection limits were 100 ng/litre (water), 10 ng/g (sediment), and 100 ng/g (fish); ND = not detected; NR = not reported

Table 10. Concentrations of TPP near sites producing or using trialkyl/aryl phosphates

Year	Location	Sample	Concentration	Number of sample (detected/analysed)	Reference
1975	New Orleans (USA)	finished water	120 ng/litre	NR	Boethling & Cooper (1985)
1976	Delaware River (USA)	river water (winter)	100-300 ng/litre	(2/5)	Sheldon & Hites (1978)
		river water (summer)	100-400 ng/litre	(11/11)	
1977	Delaware River (USA)	influent of sewage treatment plant	16 000 ng/litre	NR	Sheldon & Hites (1979)
		effluent of sewage treatment plant	2000 ng/litre	NR	
		river water	200-300 ng/litre	(3/3)	
		effluent of water treatment plant	30 ng/litre	NR	
1978	Kanawha River (USA)	river water	300-1200 ng/litre	NR	Boethling & Cooper (1985)
1980	FMC Corp. Plant (USA)	waste water	740 000 ng/litre	NR	Boethling & Cooper (1985)
		effluent water	7000 ng/litre	NR	
1980	Automobile manufacture (USA)	workplace air	0.008-0.057 mg/m^3	(6/6)	US NIOSH (1980)
1983	Saginaw River (USA)	river water	700 ng/g	(1/4)	Boethling & Cooper (1985)
NR	TPP manufacturing plant (USA)	workplace air	0.5-29.6 mg/m^3	(78/78)	Sutton et al. (1960)
NR	Waukegan Harbor, Illinois (USA)	fish (carp, goldfish)	60-150 ng/g	(3/3)	Lombardo & Egry (1979)

collected in winter, and 100-400 ng/litre in 11 out of 12 samples collected in summer. The highest level (16 000 ng/litre) of TPP was found in a waste stream entering Philadelphia's North-East Sewage Treatment plant for industrial effluents (Sheldon & Hites, 1979). Concentrations of TPP in four samples of Kanawha River (USA) water collected 13 km downstream from the outfall of an aryl phosphate manufacturing plant ranged from 300 to 1200 ng/litre (Boethling & Cooper, 1985). Mayer et al. (1981) also detected TPP (100-7900 ng/litre) in Mississippi River (USA) water sampled at St. Louis (Missouri), where two phosphate ester hydraulic fluids were being produced by Monsanto Co.

5.1.3 Sediment and soil

Relatively high concentrations of TPP have occasionally been found in sediments collected near heavily industrialized areas. Mayer et al. (1981) found TPP levels of 1000-4000 ng/g in sediment from the Saginaw River (Lake Huron) sampled at 1.6-3.2 km downstream from several plants manufacturing automobile spare parts (Boethling & Cooper, 1985). They also detected TPP levels of 10-200 ng/g at Waukegan Harbor (Illinois), Upper Saginaw River (Michigan), the Mississippi River at St. Louis (Missouri), and the Kanawha River at Winfield (West Virginia) (Mayer et al., 1981). According to the annual reports of the Environment Agency of Japan, TPP has not been found at any sampling points in Japan. The detection limits varied from 2 to 50 ng/g at the various laboratories (EAJ, 1977). Wakabayashi (1980) detected TPP levels of 0.7-3.3 ng/g in 10 out of 15 river sediment samples, and 0.2-0.3 ng/g in 2 out of 3 sea sediment samples analysed in Tokyo.

Caines & Holden (1976) identified TPP in agricultural soils collected from vineyards in Scotland, but the concentration was not reported.

5.1.4 Fish

Lombardo & Egry (1979) found TPP levels of 60-150 ng/g in carp and goldfish caught near a site at Waukegan Harbor (USA) where aryl phosphate hydraulic fluids were used. Mayer et al. (1981) detected concentrations of 100-600 ng per g in 16 out of 82 samples collected in several rivers in the USA. According to the annual reports of the Environment Agency of Japan, TPP has not been detected in fish caught at any sampling points in Japan. The detection limits ranged from 5 to 50 ng/g at the various laboratories (EAJ, 1977, 1981). Kenmotsu et al. (1981a) found TPP levels of 2-6 ng/g in 12 out of 41 samples collected from Seto Inland Sea, Japan.

5.2 General population exposure

5.2.1 Food

Gilbert et al. (1986) analysed composite total-diet samples (representative of 15 different commodity food types encompassing an average adult diet for each of eight regions in the United Kingdom) for the presence of trialkyl and triaryl phosphates. Of the food groups, offal, other animal products, and nuts consistently contained the highest levels, but the proportion of individual compounds in the different food groups varied. Trioctyl phosphate was the major component in the carcass meat, offal, and poultry groups, and there were significant amounts of TPP and TBP. Total phosphate intake was estimated to be between 0.07 and 0.1 mg per person per day.

Gunderson (1988) reported the presence of TPP in samples collected between April 1982 and April 1984 during FDA total-diet studies. The mean daily intakes of TPP were 0.3-4.4, 1.2-1.6, and 0.5-1.6 ng/kg body weight per day for infants, toddlers, and adults, respectively.

5.2.2 Drinking-water

Lebel et al. (1981) analysed TAPs in drinking-water sampled from eastern Ontario water treatment plants and found TPP levels of 0.3-2.6 ng/litre in all of the 12 samples collected. An extended survey of drinking-water was conducted in Canada (Williams & Lebel, 1981). TPP was detected at levels of 0.3-8.6 ng/litre in 7 out of 60 samples of treated potable water obtained at the treatment plants of 29 municipalities. Higher levels of TPP were present in treated water obtained from river sources, compared with samples from lake sources, and TPP was not found in potable water from wells. TPP was also detected in 11 out of 12 samples of drinking-water obtained from 12 water treatment plants located around the Great Lakes (USA and Canada) at concentrations from 0.2 to 4.8 ng/litre (Williams et al., 1982). Sheldon & Hites (1979) reported a relatively high level of TPP (30 ng/litre) in finished drinking-water sampled from a water treatment plant located near a sewage treatment plant handling industrial effluents.

In general, the concentration of TPP in drinking-water is 100- to 1000-fold lower than that in river or lake water. Due to its adsorption to sediment, TPP can efficiently be removed by filtration at water treatment plants.

5.2.3 Human tissues

There has been only one report of TAPs being present in human adipose tissues (Lebel & Williams, 1983). TPP was not detected.

5.3 Occupational exposure

Sutton et al. (1960) investigated the concentration of TPP in the air of a TPP-manufacturing plant. The levels found at various locations are indicated in Table 11.

Table 11. Air concentrations of TPP at various locations in a manufacturing plant

Location	Number of samples	Range (mg/m^3)	Mean (mg/m^3)
Reactor room	6	1.4-2.2	1.8
Purification area			
General room air	6	1.0-3.9	2.4
Receiving tank	10	5.0-29.6	12.0
Flaker room			
General room air	6	1.8-3.7	2.6
Flaker	11	2.6-6.8	4.5
Bagging area			
General room air	5	2.2-7.8	4.1
Bagger	23	0.5-20.8	8.2
Stacking bags	11	0.7-7.4	5.4

TPP has been detected at concentrations of 0.008-0.057 mg/m^3 in the air at automobile manufacturing plants where hydraulic fluids are used (US NIOSH, 1980).

6. EFFECTS ON ORGANISMS IN THE ENVIRONMENT

Summary

The primary productivity of green algal cultures was inhibited (to 50%) by exposure to TPP (0.26 to 0.5 mg/litre) for 7 days, and the nitrogenase activity of cyanobacteria (blue-green algae) was inhibited at 5 mg/litre. Fungal spore germination was unaffected at 5×10^{-3} mol/litre.

The 48-h LC_{50} for Daphnia is 1.0 mg/litre and 96-h LC_{50} values for fish range from 0.36 to 290 mg/litre. The no-observed-effect level for growth and survival of rainbow trout fry is 1.4 µg/litre.

There is no information on the toxicity of TPP to organisms living in or ingesting sediment and none on terrestrial species other than fungi.

6.1 Unicellular algae and fungi

TPP was the most toxic compound among six TAPs tested for effects on the primary productivity of algae (Wong & Chau, 1984). The growth of green algae was completely inhibited at concentrations of 1 mg/litre or more but stimulated at lower concentrations (0.1 and 0.05 mg/litre) (Wong & Chau, 1984). The nitrogenase activity, measured by the acetylene reduction technique, of a cyanobacterium (blue-green alga: *Anabaena flos-aquae*) was affected by TPP. Additions of 0.1, 1.0, and 5.0 mg/litre reduced the nitrogenase activity to 84, 77, and 68% of the control value, respectively (Wong & Chau, 1984). These data are summarized in Table 12.

TPP did not show any toxicity, as measured by spore germination, to the fungus *Aspergillus niger* at a concentration of 5×10^{-3} mol/litre. However, it inhibited fungal respiration by 8-9% (Eto et al., 1975).

6.2 Aquatic organisms

Data on the toxicity of TPP to aquatic organisms are given in Table 13.

Table 12. Toxicity of TPP and its products for freshwater unicellular algae

Organism	Chemical	Temperature (°C)	Species	Effect	Concentration (mg/litre)	Reference
Alga	TPP	20	*Ankistrodesmus falcatus acicularis*	7-day IC_{50} for primary productivity	0.26	Wong & Chau (1984)
Green alga	TPP	20	*Scenedesmus quadricaudata*	7-day IC_{50} for primary productivity	0.50	Wong & Chau (1984)
Green alga	TPP P50E P115E		*Selenastrum capricornutum* *Selenastrum capricornutum* *Selenastrum capricornutum*	96-h EC_{50} 96-h EC_{50} 96-h EC_{50}	2 5 > 1000	Mayer et al. (1981) Mayer et al. (1981) Mayer et al. (1981)
Cyanobacterium (blue-green alga)	TPP	20	*Anabaena flos-aquae* 61% of control	28-h Inhibition of acetylene reduction	5	Wong & Chau (1984)
Lake Ontario phytoplankton	TPP	20		IC_{50} for primary productivity	0.2	Wong & Chau (1984)

Table 13. Toxicity of TPP for aquatic organisms

Organisms	Age/size	Temperature (°C)	pH	Flow/stat	Hardness (mg/litre)	End-point or criteria used	Parameter	Concentration (mg/litre)	Reference
Rainbow trout (*Salmo gairdneri*)	Fry: 0.11 g, 24 mm, 12 days past swim-up stage			stat			96-h LC_{50}	0.36	Palawski et al. (1983)
	Fry: 0.11 g, 24 mm, 12 days past swim-up stage			stat		immobility, mortality, loss of equilibrium	96-h LC_{50}	0.30	Palawski et al. (1983)
	Fry	12(±1)	7.2	stat	272	mortality and growth inhibition	96-h LC_{50}	0.40	Mayer et al. (1981)
				flow	272		90-d EC_0	>0.0014	Mayer et al. (1981)
	0.60 g	12	7.4	stat	40		96-h LC_{50}	0.37	Mayer & Ellersieck (1986)
Fathead minnow (*Pimephales promelas*)	Egg and fry			stat		mortality and growth inhibition	96-h LC_{50}	0.66	Mayer et al. (1981)
	Egg and fry			flow			90-d EC_0	0.087-0.27	Mayer et al. (1981)
				flow			90-d EC_0	>0.23	Mayer et al. (1981)
	1.00 g	22	7.3	stat	44		96-h LC_{50}	1.0	Mayer & Ellersieck (1986)
Sheepshead minnow (*Cyprinodon viriegatus*)				stat			96-h LC_{50}	0.32-0.56	Mayer et al. (1981)

Table 13 (contd).

Species	Size	Temperature (°C)	pH	Test type	Hardness	Endpoint	Value (mg/litre)	Reference
Bluegill (*Leptomis macrochirus*)	33-75 mm	23	7.6-7.9	stat	55	96-h LC$_{50}$	290	Dawson et al. (1977)
Killifish (*Oryzias latipes*)	0.1-0.2 g	25		stat		96-h LC$_{50}$	1.2	Sasaki et al. (1981)
Goldfish (*Carassius auratus*)	0.8-2.8 g			stat		96-h LC$_{50}$	0.70	Sasaki et al. (1981)
Channel catfish (*Ictalurus punctatus*)	0.23 g	22	7.5	stat	38	96-h LC$_{50}$	0.42	Mayer & Ellersieck (1986)
Tidewater silverside (*Menidia beryllina*)	40-100 mm	20		stat		96-h LC$_{50}$	95	Dawson et al. (1977)
Mysid shrimp (*Mysidopsis bahia*)				stat		96-h LC$_{50}$	0.18-0.32	Mayer et al. (1981)
Water flea (*Daphnia magna*)				stat		48-h EC$_{50}$	1.0	Mayer et al. (1981)

The 96-h LC_{50} values for pure TPP to fish range from 0.36 mg/litre for the rainbow trout (Palawski et al., 1983) to 290 mg/litre for the bluegill (Dawson et al., 1977).

The growth and survival of rainbow trout fry were not affected when they were exposed to TPP at a concentration of 0.0014 mg/litre (Mayer et al., 1981). At 0.23 mg/litre, the survival of fathead minnow fry was significantly reduced, but neither the growth of the survivors nor hatchability was affected (Mayer et al., 1981; Palawski et al., 1983).

Sublethal effects of TPP on fish include morphological and behavioural abnormalities (Wagemann et al., 1974; Lockhart et al., 1975). Spinal curvature was observed in surviving rainbow trout exposed for 24-72 h at concentrations near the LC_{50} (Sasaki et al., 1981; Palawski et al., 1983).

Death of goldfish occurred in a 20-litre water tank in which a piece (18 x 38 cm) of car seat upholstery containing TPP had been immersed (Ahrens et al., 1978). Goldfish exposed to TPP (concentration not stated) showed histopathological lesions characterized by congestion, degeneration, and haemorrhage of the smaller blood vessels, principally venules and capillaries. Such vascular pathology was most pronounced in the gills. Similar but less pronounced congestion of the smaller blood vessels was noted in the brain, spinal cord, pseudobranch and kidneys (Ahrens et al., 1978). Immobility of fish exposed to 0.21-0.29 mg TPP/litre disappeared within 7 days after exposure had stopped (Palawski, et al., 1983).

Exposure of aquatic organisms to TPP would normally arise from spillage of hydraulic fluids containing this compound. Studies have been made of the effects of various products, especially Pydraul 50E and 115E, Houghtosafe 1120, and Santicizer 154, on fish and aquatic invertebrates. Where comparisons have been made between different components of the fluids, TPP has been shown to be responsible for most of the acute toxicity observed. However, certain characteristic sublethal symptoms seen with these hydraulic fluids (such as cataracts of the eye lens, effects on bone development and collagen content, haemorrhagic lesions of the dorsal and gill regions, and vertebral deformity) do not occur after exposure to TPP. They are therefore caused by other fluid components (Wagemann et al., 1974; Dawson et al., 1977; Nevins & Johnson, 1978; Mayer et al., 1981; Adams et al., 1983).

6.3 Insects

In studies on 5th-instar small brown planthopper larvae *(Laodelphax striatellus)*, the chemical being applied by contact, the 21-h LD_{50} was 570.2 µg TPP per tube, but TPP-OH was without effect (Eto et al., 1975). The 24-h LD_{50} in similar studies on adult female (2-5 days old) house flies *(Musca domestica)* was >1000 µg TPP per jar (Plapp & Tong, 1966). When adult female (4-5 days old) green rice leafhoppers were treated topically, the 24-h LD_{50} for TPP was 4.6 mg/g and for TPP-OH was 11.53 mg/g (Eto et al., 1975).

7. KINETICS AND METABOLISM

No data on the kinetics and metabolism of TPP in experimental animals are available. Eto et al. (1975) reported that treated houseflies transform TPP into diphenyl *p*-hydroxyphenyl phosphate (TPP-OH) *in vivo*.

8. EFFECTS ON ANIMALS AND *IN VITRO* SYSTEMS

Summary

Acute toxicity data exist for several species of animals and indicate low toxicity via the oral and dermal routes (1320 to 10 800 mg/kg and >7900 mg/kg, respectively). No inhalation data are available. TPP also exhibits low toxicity in short-term studies and is not irritant to mouse skin. In rats, no effects were seen in mothers or offspring following repeated dietary exposure of 166-690 mg/kg per day for a period of 91 days, including mating and gestation periods.

The neurotoxicity of TPP has been debated since the early studies of Smith et al. (1930, 1932), which reported delayed neuropathy in cats and monkeys exposed to TPP in acute and short-term studies. However, Wills et al. (1979) could demonstrate no ataxia or neuropathic damage in cats exposed to 99.9%-pure TPP. Consequently, the validity of the Smith studies has been questioned. Other toxicity studies using behavioural and morphological end-points have demonstrated that TPP administered short-term to cats and chickens fails to produce neurotoxic changes. A mixture of triaryl (including cresyl and phenyl) phosphates produced neurochemical changes and minor peripheral nerve pathology in the caudal nerve of rats; acute intraperitoneal injection of 150 mg or less produced neither biochemical nor morphological change.

Negative results have been reported for several in vitro mutagenicity studies. No satisfactory studies are available on carcinogenicity.

8.1 Single exposure

Acute toxicity data resulting from single exposure to TPP are summarized in Table 14. Little information is available on the acute signs of toxicity.

8.2 Short-term exposure

Sutton et al. (1960) reported a 35-day feeding study in male Holtzman rats with TPP at doses of 1 and 5 g/kg. Depression of body weight gain and an increase of liver

Table 14. Acute toxicity of triphenyl phosphate

Species	Route of administration	LD_{50} (mg/kg)	Reference
Rat	oral	3500	Hierholzer et al. (1957)
Rat	oral	3800	Antonyuk (1974)
Rat	oral	10 800	Johannsen et al. (1977)
Rat	oral	> 5000	US EPA (1986)
Mouse	oral	> 3000	Sutton et al. (1960)
Mouse	oral	1320	Antonyuk (1974)
Mouse	oral	> 5000	US EPA (1986)
Guinea-pig	oral	> 4000	Sutton et al. (1960)
Chicken	oral	> 2000	Smith et al. (1932)
Chicken	oral	> 5000	Johannsen et al. (1977)
Rabbit	dermal	> 7900	Johannsen et al. (1977)

weight were observed in the high-dose group. No haematological changes were found.

In studies by Hinton et al. (1987), TPP was fed to weanling Sprague-Dawley rats (10 of each sex per group) at dose levels of 0, 2.5, 5, 7.5, or 10 g/kg for 120 days. The immunotoxicity evaluation included total protein analysis, electrophoretic analysis of serum proteins, lymphoid organ weights in relation to growth, and histopathology, together with expanded immunohistochemical evaluation of B- and T-lymphocyte regions in the spleen, thymus, and lymph nodes using immunoperoxidase staining. Assessment was made of the humoral response to a T-lymphocyte-dependent antigen, sheep red blood cells; it began at mid-term of the feeding period for the primary response and was followed by secondary and tertiary booster immunizations at intervals of 3 weeks. The kinetics of the response were measured by haemolysin assay of relative antibody titres at days 3, 4, 5, and 6 post injection. No significant effects on the response were noted for either sex at any of the dose levels tested. The only effects noted were a decreased rate of growth at high levels of TPP and increased levels of α- and β-globulins (Hinton et al., 1987).

When Antonyuk (1974) administered TPP orally for 3 months to rats at doses of 380 or 1900 mg/kg, there were no deaths, no abnormal growths, and no inhibition of cholinesterase activity. In another study, Antonyuk (1974) administered 650 to 1900 mg/kg orally to rats for 3 months with no significant toxic effects.

8.3 Skin irritation

No significant skin irritation was observed when a gauze pad soaked with approximately 0.5 ml of a 70% solution of TPP in alcohol was applied to the skin of mice for 72 h (Sutton et al., 1960).

8.4 Reproduction

In studies by Welsh et al. (1987), male and female Sprague-Dawley (Spartan) rats (40 of each sex per group) were fed dietary levels of 0, 2.5, 5, 7.5, or 10 mg TPP/kg (from 4 weeks post weaning for 91 days, through mating and gestation). At these dietary levels, the daily intake of TPP during pregnancy was 0, 166, 341, 516, and 690 mg/kg body weight, respectively. TPP exposure had no toxic effects on mothers or offspring at these dosages. The types of developmental anomalies were similar in both treated and control animals, and no significant increase in the incidence of anomalies was seen in the treated groups as compared to control values. TPP was not teratogenic in Sprague-Dawley rats at the levels tested.

8.5 Mutagenicity

Szybalski (1958) reported negative results with TPP in a paper disk method using streptomycin-dependent mutants of *E.coli*.

TPP did not demonstrate mutagenic activity in microbial assays employing *Salmonella typhimurium* (TA 1535, TA 1537, TA 1538, TA 98, and TA 100 strains) and *Saccharomyces cerevisiae* (D4 strain) indicator organisms. All studies were carried out both in the presence and absence of metabolic activation (Monsanto, 1979).

Negative results were also reported in Ames tests conducted with *Salmonella typhimurium* strains TA 98, TA 100, TA 1535, and TA 1537, in the absence or presence of rat liver S9 (Zeiger et al., 1987).

TPP was tested for its ability to induce mutations at the thymidine kinase (TK) locus in cultured L5178Y mouse lymphoma cells. When tested with or without metabolic activation, TPP did not induce significant mutations at the TK locus (Monsanto, 1979).

8.6 Carcinogenicity

Theiss et al. (1977) studied the occurrence of lung adenomas in strain A/St male mice, 6 to 8 weeks old, using doses of 80, 40, or 20 mg TPP/kg injected intraperitoneally 1, 3, and 18 times, respectively, into groups of 20 mice. Twenty-four weeks after the first injection, the animals were sacrificed, and the frequency of lung tumours was compared with that in the control group of 50 animals treated with tricarpylin (vehicle). The pulmonary adenoma response to TPP was not significantly greater than the response of the control mice. This study was considered inadequate due to the low survival of animals in two of the three experimental groups and the short duration of the study.

8.7 Neurotoxicity

In 1930, Smith and his associates found that single and multiple doses of technical grade TPP produced generalized delayed paralysis in cats and monkeys but not in chickens or rabbits (Smith et al., 1930).

Smith et al. (1932) attempted to ascertain the minimum lethal dose of TPP. Rabbits survived after an intramuscular injection of 1 g/kg; chickens also were unaffected after oral administrations of 0.5 to 2 g/kg. In cats, the minimum toxic dose by subcutaneous injection was about 0.2 g/kg and the reaction was of the delayed type; a neurotoxic action and flaccid paralysis were followed by death.

Johannsen et al. (1977) dosed chickens with cumulative doses of 60 g/kg but failed to produce ataxia or neuropathology suggestive of organophosphate-induced delayed neuropathy (OPIDN).

In an attempt to re-evaluate the delayed neurotoxicity, Wills et al. (1979) reported that 99.9%-pure TPP did not produce any evidence of axonal degeneration, demyelination, or any other pathological changes at 11 levels of the nervous system (from the cerebral cortex to peripheral nerves) when subcutaneously injected into cats at doses of 0.4, 0.7, or 1.0 g/kg. Prostration occurred at the higher doses. Wills et al. (1979) suggested that the samples of TPP used by Smith et al. (1930) may have contained impurities that were capable of producing axonal degeneration and demyelination.

Sobotka et al. (1986) fed young male Sprague-Dawley rats (10 per group) diets containing TPP at levels of 0, 2.5, 5, 7.5, or 10 g/kg, for 4 months. Treatment-related decreases in growth rate, in the absence of changes in food consumption, were found at all dietary levels above 2.5 g/kg. There was no evidence of neuromotor toxicity following subchronic dietary exposure to TPP.

In a study by Vainiotalo et al. (1987), a commercial cresyl diphenyl phosphate preparation was analysed and found to contain approximately 35% triphenyl phosphate, 45% cresyl diphenyl phosphates, 18% dicresyl phenyl phosphates, and 2% tricresyl phosphates. The product was almost free of the o-cresyl isomers, as revealed by the analysis of its alkaline hydrolysis products. A single intraperitoneal injection (150 or 300 mg/kg) of this mixture caused the induction of microsomal cytochrome P-450 in the liver of Wistar rats, a concomitant increase in the activities of mixed function monooxygenases, and proliferation of smooth endoplasmic reticulum 24 h after the treatment. The activity of pseudocholinesterase in blood was inhibited 4 h and 24 h after the injection but the effect leveled off. Treatment with 300 mg/kg inhibited brain $2',3'$-cyclic-nucleotide $3'$-phosphohydrolase through the 2-week observation period associated with demyelination in peripheral nerves.

8.8 In Vitro studies

In vitro TPP was found to cause significant direct inhibition of monocyte antigen presentation at non-cytotoxic concentrations as low as 1 μmol/litre (Esa et al., 1988).

Effects on Animals and In Vitro Systems

Mochida et al. (1988) studied the *in vitro* cytotoxicity of TPP to human, monkey, and dog cell lines and demonstrated a dose-dependent inhibition of growth.

9. EFFECTS ON HUMANS

Sutton et al. (1960) found no evidence of neurological disease or other abnormalities in 32 workers exposed to TPP vapour, mist, or dust (at a time-weighted air concentration of 3.5 mg/m^3) for an average of 7.4 years. In six of these workers, who were exposed more regularly to TPP, there was a statistically significant asymptomatic reduction in erythrocyte cholinesterase values, but no plasma cholinesterase depression.

In 39 workers exposed to an organophosphate ester mixture with about 30% TPP and 70% different isopropyl TPPs, a significantly lower level of serum IgM and a lower activity (of borderline significance; $p = 0.05$) of erythrocyte cholinesterase, compared to controls, were reported. However, plasma cholinesterase activity and the other observed parameters were not significantly affected (Emmett et al., 1984).

A few individuals have been reported to show positive reactions in patch tests using cellulose acetate film containing both TCP and TPP. However, the causative agent could not be identified and may have been TCP (Hjorth, 1964). A single case of allergy to spectacle frames could also have been due to TCP (Carlsen et al., 1986).

10. EVALUATION OF HUMAN HEALTH RISKS AND EFFECTS ON THE ENVIRONMENT

10.1 Evaluation of human health risks

Animal data indicate that TPP has low toxicity. It produces no irritant effect on animal skin. Despite an early report to the contrary, TPP is not considered neurotoxic in animals or man. The no-observed-adverse-effect level on mothers and offspring from a 90-day rat study was at 690 mg/kg per day. Both exposure of the general population and occupational exposure to TPP are low.

TPP is not mutagenic.

The available data indicate no hazard to humans.

10.1.1 Exposure levels

Exposure of the general population to TPP through various environmental media, including drinking-water, is likely. The concentrations of TPP measured in drinking-water in Canada and the USA are extremely low. TPP has often been detected in urban air, although the levels are low. Vaporization of TPP from heated vinyl automotive upholstery under hot weather has been suggested, but no data on concentrations in cars are available. In a survey of TAPs in human adipose tissues, TPP was not detected. There are insufficient data to evaluate the significance of the general population exposure to TPP.

Significant air concentrations (0.5-29.6 mg/m^3) have been reported in a TPP manufacturing plant, but recent figures are not available. More data on occupational exposure to TPP in manufacturing plants are required.

10.1.2 Toxic effects

The toxicity profile of TPP is quite inadequate for a full evaluation of its hazard.

There is no evidence that TPP has mutagenic activity in bacteria or that it has carcinogenic activity, based on a study in one animal species. No evidence that TPP causes delayed neurotoxicity has so far been obtained in animal experiments. In a 35-day feeding study in rats, depression of body weight gain and increase in liver weight were observed at a dose of 5 g/kg. No adequate data are available on the effects of TPP on reproduction, i.e. function of gonads, fertility, parturition, and growth and development of offspring.

Contact dermatitis due to TPP has been described.

10.2 Evaluation of effects on the environment

Water concentrations of TPP in the environment are low and toxic effects on aquatic organisms are unlikely. Spills of hydraulic fluids containing TPP would be expected to cause local kills. Since TPP is removed rapidly from the tissues of fish when exposure ends and bioconcentration factors are moderate, bioaccumulation is not considered to be a hazard.

High concentrations of TPP in sediment near production plants have been reported. TPP bound to sediment has been shown to be bioavailable to one organism living in sediment, but no toxicity data on sediment-living or sediment-ingesting species exist. There is, therefore, the possibility of effects on aquatic communities.

10.2.1 Exposure levels

TPP is found in air, surface water, soil, sediment, and aquatic organisms sampled in heavily industrialized areas. The highest reported concentration of TPP in industrial water effluent is 16 μg/litre, while that in river water is 7.9 μg/litre. Taking into account the rapid biodegradation of TPP in aqueous environments, normal concentrations of TPP in aqueous environments are unlikely to adversely affect aquatic organisms. However, disposal of TPP-treated vinyl fabric upholstery into a pond would result in a sufficiently high concentration of TPP to kill fish.

10.2.2 Toxic effects

Among the various triaryl phosphates, TPP is the most acutely toxic compound to fish, shrimps, and daphnids. The

96-h LC_{50} of TPP for fish ranges from 0.36 mg/litre in rainbow trout to 290 mg/litre in bluegill. Salmonids are generally sensitive to TPP, but the growth and survival of rainbow trout fry were not affected when they were exposed to TPP at a concentration of 0.0014 mg/litre. Histopathological lesions in goldfish exposed to TPP consist of congestion, degeneration, and haemorrhage of the small blood vessels, principally venules and capillaries. Such vascular pathology is most pronounced in the gills.

The growth of algae was completely inhibited at TPP concentrations of 1 mg/litre or more but was stimulated at lower concentrations (0.1 and 0.05 mg/litre). The nitrogenase activity of *Anabaena flos-aquae* was significantly reduced, even at 0.1 mg/litre.

11. RECOMMENDATIONS

11.1 Recommendations for further research

There is a need for skin sensitization, *in vitro* cytogenicity, and pharmacokinetic studies on triphenyl phosphate.

REFERENCES

ADAMS, W.J., KIMERLE, R.A., HEIDOLPH, B.B., & MICHAEL, P.R. (1983) *Field comparison of laboratory-derived acute and chronic toxicity data,* Philadelphia, American Society of Testing and Materials, pp. 367-385 (ASTM STP 802).

AHRENS, V.D., HENION, J.D., MAYLIN, G.A., LEIBOVITZ, L., ST. JOHN, L.E., Jr, & LISK, D.J. (1978) A water-extractable toxic compound in vinyl upholstery fabric. *Bull. environ. Contam. Toxicol.,* 20: 418-422.

ALDOUS, C.N. (1982) Alterations in rat brain norepinephrine and dopamine levels and synthesis rates in response to five neurotoxic chemicals: acrylamide, 2,5-hexanedione, tri-o-tolyl phosphate, leptophos, and methyl mercuric chloride. *Diss. Abstr. Int.,* 43(5): 1441B.

ANTONYUK, O.K. (1974) Hygienic evaluation of the plasticizer triphenyl phosphate added to polymer composition. *Gig. i Sanit.,* 8: 98-99.

BARNARD, P.W.C., BUNTON, C.A., LLEWELLYN, D.R., VERNON, C.A., & WELCH, V.A. (1961) The reactions of organic phosphates. Part V. The hydrolysis of triphenyl and trimethyl phosphates. *J. Chem. Soc.,* 1961: 2670-2676.

BARNARD, P.W.C., BUNTON, C.A., KELLERMAN, D., MHALA, M.M., SILVER, B., VERNON, C.A., & WELCH, V.A. (1966) Reactions of organic phosphates. Part VI. The hydrolysis of aryl phosphates. *J. Chem. Soc. (B),* 1966: 227-235.

BARRETT, H., BUTLER, R., & WILSON, I.B. (1969) Evidence for a phosphorylenzyme intermediate in alkaline phosphatase catalyzed reactions. *Biochemistry,* 8: 1042-1047.

BHATTACHARYYA, J., BHATTACHARYYA, K., SENGUPTA, P.K., & GANGULY, S.K. (1974) Detection and estimation of tricresyl phosphate in mustard oil. *Forensic Sci.,* 3: 263-270.

BLOOM, P.J. (1973) Application des chromatographies sur couch mince et gazliquide à l'analyse qualitative et quantitative des esters des acides phosphorique et phosphoreux. *J. Chromatogr.,* 75: 261-269.

BOETHLING, R.S. & COOPER, J.C. (1985) Environmental fate and effects of triaryl and trialkyl/aryl phosphate esters. *Residue Rev.,* 94: 49-99.

BOWERS, W.D., PARSONS, M.L., CLEMENT, R.E., EICEMAN, G.A., & KARASEK, F.W. (1981) Trace impurities in solvents commonly used for gas chromatographic analysis of environmental samples. *J. Chromatogr.,* 206: 279-288.

BRAUN, D. (1965) [Qualitative analysis of plasticizers by using thin-layer chromatography.] *Chimia,* 19(2): 77-82 (in German).

CAINES, L.A. & HOLDEN, A.V. (1976) Stream pollution by an organomercury compound. *Bull. environ. Contam. Toxicol.,* 16(4): 483-485.

CARLSEN, L., ANDERSEN, K.E., & EGSGAARD, H. (1986) Triphenyl phosphate allergy from spectacle frames. *Contact dermatitis*, 15: 274-277.

Ciba-Geigy Corporation: Acute oral toxicity to rats and mice, EPA-OTS document 86-870000069.

DAFT, J.L. (1982) Identification of aryl/alkyl phosphate residues in foods. *Bull. environ. Contam. Toxicol.*, 29: 221-227.

DAWSON, G.W., JENNINGS, A.L., DROZDOWSKI, D., & RIDER, E. (1977) The acute toxicity of 47 industrial chemicals to fresh and saltwater fishes. *J. hazard. Mater.*, 1(4): 303-318.

DEO, P.G. & HOWARD, P.H. (1978) Combined gas-liquid chromatographic mass spectrometric analysis of some commercial aryl phosphate oils. *J. Assoc. Off. Anal. Chem.*, 61: 266-271.

DRUYAN, E.A. (1975) [Separation and determination of tricresyl phosphate, triphenyl phosphate, phenol, o-, m-, and p-cresol by thin-layer chromatography.] *Gig. i Sanit.*, 10: 62-65 (in Russian).

EAJ (1977) [Environmental monitoring of chemicals.] Tokyo, Environment Agency Japan, pp. 212-214 (Environmental Survey Report Series No. 3) (in Japanese).

EAJ (1981) Environmental monitoring of chemicals: Environmental survey report of 1978, 1979 F.Y., Tokyo, Environment Agency Japan, pp. 171-183.

EMMETT, E.A., DANNENBERG, A.M., LEWIS, P.G., FOX, R., BLEECKER, M., TANAKA, F., SYNKOWSKI, D.R., & LEVINE, M.S. (1984) A clinical study of industrial workers exposed to organophosphate esters and a comparison group, Pittsfield, Massachusetts, General Electric Company (Prepared for the US Environmental Protection Agency, Washington).

EMMETT, E.A., LEWIS, P.G., TANAKA, F., BLEECKER, M., FOX, R., DARLINGTON, A.C., SYNKOWSKI, D.R., DANNENBERG, A.M., Jr, TAYLOR, W.J., & LEVINE, M.S. (1985) Industrial exposure to organophosphorus compounds. Studies of a group of workers with a decrease in esterase-staining monocytes. *J. occup. Med.*, 27(12): 905-914.

ESA, A.H., WARR, G.A., & NEWCOMBE, D.S. (1988) Immunotoxicity of organophosphorous compounds. *Clin. Immunol. Immunopathol.*, 49: 41-52.

ETO, M., HASHIMOTO, Y., OZAKI, K., & SASAKI, Y. (1975) [Fungitoxicity and insecticide synergism of monothioquinol phosphate esters and related compounds.] *Botyu Kagaku*, 40: 110-117 (in Japanese).

FINNEGAN, R.A. & MATSON, J.A. (1972) Irradiation of triaryl phosphate esters. A new photochemical coupling reaction. *J. Am. Chem. Soc.*, 94: 4780-4782.

FUKUSHIMA, M. & KAWAI, S. (1986) [Present status and transition of selected organophosphoric acid triesters in the water area of Osaka city.] *Seitai Kagaku*, 8: 13-24 (in Japanese).

References

GILBERT, J., SHEPHERD, M.J., WALLWORK, M.A., & SHARMAN, M. (1986) A survey of trialkyl and triaryl phosphates in the United Kingdom total diet samples. *Food Addit. Contam.*, 3(2): 113-122.

GUNDERSON, E.L. (1988) FDA total diet study, April 1982 - April 1984, dietary intakes of pesticides, selected elements and other chemicals. *J. Assoc. Off. Anal. Chem.*, 71(6): 1200-1209.

HATTORI, Y., ISHITANI, H., KUGE, Y., & NAKAMOTO, M. (1981) [Environmental fate of organic phosphate esters.] *Suishitu Odaku Kenkyu*, 4: 137-141 (in Japanese).

HIERHOLZER, K., NOETZEL, H., & SCHMIDT, L. (1957) [Comparative toxicological study on triphenyl phosphate and tricresyl phosphate.] *Arzneimittelforschung*, 7: 585-588.

HINE, C., ROWE, V.K., WHITE, E.R., DARMER, K.I., Jr, & YOUNGBLOOD, G.T. (1981) In: Clayton, G.D. & Clayton, F.E., ed. Patty's industrial hygiene and toxicology, 3rd revised ed., New York, Wiley-Interscience, Vol. 2A, pp. 2362-2363.

HINTON, D.M., JESSOP, J.J., ARNOLD, A., ALBERT, R.H., & HINES, F.A. (1987) Evaluation of immunotoxicity in a subchronic feeding study of triphenyl phosphate. *Toxicol. ind. Health*, 3(1): 71-89.

HJORTH, N. (1964) Contact dermatitis from cellulose acetate film. Cross-sensitization between tricresyl phosphate (TCP) and triphenyl phosphate (TPP). *Contact dermatitis*, 12: 86-100.

HOLLIFIELD, H.C. (1979) Rapid nephelometric estimate of water solubility of highly insoluble organic chemicals of environmental interest. *Bull. environ. Contam. Toxicol.*, 23: 579-586.

HOWARD, P.H. & DEO, P.G. (1979) Degradation of aryl phosphates in aquatic environments. *Bull. environ. Contam. Toxicol.*, 22: 337-344.

HUDEC, T., THEAN, J., KUEHL, D., & DOUDHERTY, R.C. (1981) Tris(dichloropropyl)phosphate, a mutagenic flame retardant: Frequent occurrence in human seminal plasma. *Science*, 211: 951-952.

ISHIKAWA, S., TAKETOMI, M., & SHINOHARA, R. (1985) Determination of trialkyl and triaryl phosphates in environmental samples. *Water Res.*, 19: 119-125.

JOHANNSEN, F.R., WRIGHT, P.L., GORDON, D.E., LEVINSKAS, G.J., RADUE, R.W., & GRAHAM, P.R. (1977) Evaluation of delayed neurotoxicity and dose-response relationships of phosphate esters in the adult hen. *Toxicol. appl. Pharmacol.*, 41: 291-304.

KANAZAWA, J. (1978) Studies on formulation and residue analysis of pesticides. *J. Pestic. Sci.*, 3: 185-193.

KAWAI, S., FUKUSHIMA, M., ODA, K., & UNO, G. (1978) [Water pollution caused by organophosphoric compounds.] *Kankyo Gijyutsu*, 7: 668-675 (in Japanese).

KENMOTSU, K., MATSUNAGA, K., & ISHIDA, T. (1980a) [Multiresidue determination of phosphoric acid triesters in fish, sea sediment and sea water.] *J. Food Hyg. Soc. Jpn*, 21: 18-31 (in Japanese).

KENMOTSU, K., MATSUNAGA, K., & ISHIDA, T. (1980b) [Studies on the mechanisms of biological activities of various environmental pollutants. V: Environmental fate of organic phosphoric acid triesters.] *Okayama-ken Kankyo Hoken Senta Nemnpo*, 4: 103-110 (in Japanese).

KENMOTSU, K., MATSUNAGA, K., & ISHIDA, T. (1981a) [Studies on the biological toxicity of several pollutants in environments.] *Okayama-ken Kankyo Hoken Senta Nempo*, 5: 167-175 (in Japanese).

KENMOTSU, K., MATSUNAGA, K., SAITO, N., & OGINO, Y. (1981b) [An environmental survey of chemicals. XVII. Multiresidue determination of organic phosphate esters in environment samples.] *Okayama-ken Kankyo Hoken Senta Nempo*, 5: 145-156 (in Japanese).

KENMOTSU, K., MATSUNAGA, K., SAITO, N., OGINO, Y., & ISHIDA, T. (1982a) [An environmental survey of chemicals. XIX. Determination of organophosphoric acid triesters (2).] Okayama-ken Kankyo Hoken Senta Nempo, 6: 126-132 (in Japanese).

KENMOTSU, K., MATSUNAGA, K., SAITO, N., OGINO, Y., & ISHIDA, T. (1982b) [Studies on the biological toxicity of several pollutants in environments. VII. GC/MS Spectrometric determination of organophosphoric acid triesters in sediment.] *Okayama-ken Kankyo Hoken Senta Nempo*, 6: 142-152 (in Japanese).

KENMOTSU, K., NAKAGIRI, M., OGINO, Y., MATSUNAGA, K., & ISHIDA, T. (1983) [An environmental survey of chemical. XXII. GC/MS spectrometric determination of organophosphoric acid triesters in sediment (2).] *Okayama-ken Kankyo Hoken Senta Nempo*, 7: 143-149 (in Japanese).

KONASEWICH, D., TRAVERSY, W., & ZAR, H. (1978) Status report on organic and heavy metal contaminants in the Lakes Erie, Michigan, Huron and Superior basins. *Great Lakes Water Qual. Bd*.

LEBEL, G.L. & WILLIAMS, D.T. (1983) Determination of organic phosphate triesters in human adipose tissue. *J. Assoc. Off. Anal. Chem.*, 66: 691-699.

LEBEL, G.L., WILLIAMS, D.T., GRIFFITH, G., & BENOIT, F.M. (1979) Isolation and concentration of organophosphorus pesticides from drinking water at the ng/L level, using macroreticular resin. *J. Assoc. Off. Anal. Chem.*, 62: 241-249.

LEBEL, G.L., WILLIAMS, D.T., & BENOIT, F.M. (1981) Gas chromatographic determination of trialkyl/aryl phosphates in drinking water, following isolation using macroreticular resin. *J. Assoc. Off. Anal. Chem.*, 64: 991-998.

LHOMME, V., BRUNEAU, C., SOYER, N., & BRAULT, A. (1984) Thermal behavior of some organic phosphates. *Ind. Eng. Chem. Prod. Res. Dev.*, 23: 98-102.

References

LOCKHART, W.L., WAGEMANN, R., CLAYTON, J.W., GRAHAM, B., & MURRAY, D. (1975) Chronic toxicity of a synthetic triaryl phosphate oil to fish. *Environ. Physiol. Biochem.*, 5: 361-369.

LOCKHART, W.L., METNER, D.A., BLOUW, A.P., & MUIR, D.C.G. (1982) Prediction of biological availability of organic chemical pollutants to aquatic animals and plants. In: Pearson, J.G., Foster, R.B., & Bishop, W.E., ed. *Aquatic toxicology and hazard assessment: Fifth conference,* Philadelphia, American Society for Testing and Materials, pp. 259-272 (ASTM STP 766).

LOMBARDO, P. & EGRY, I.J. (1979) Identification and gas-liquid chromatographic determination of aryl phosphate residues in environmental samples. *J. Assoc. Off. Anal. Chem.*, 62(1): 47-51.

LU, P.-Y. & METCALF, R.L. (1975) Environmental fate and biodegradability of benzene derivatives as studied in a model aquatic ecosystem. *Environ. Health Perspect.*, 10: 269-284.

MAYER, F.L., Jr, MAYER, K.S., & ELLERSIECK, M.R. (1986) Relation of survival to other endpoints in chronic toxicity tests with fish. *Environ. Toxicol. Chem.*, 5(8): 737-748.

MAYER, F.L., ADAMS, W.J., FINLEY, M.T., MICHAEL, P.R., MEHRLE, P.M., & SAEGER, V.W. (1981) Phosphate ester hydraulic fluids: An aquatic environmental assessment of Pydrauls 50E and 115E, In: Branson, D.R. & Dickson, K.L., ed. *Aquatic Toxicology and Hazard Assessment: Fourth Conference,* Philadelphia, American Society for Testing and Materials, pp. 103-123 (ASTM STP 737).

MOCHIDA, K., GOMYODA, M., FUJITA, T., & YAMAGATA, K. (1988) Tricresyl phosphate and triphenyl phosphate are toxic to cultured human, monkey, and dog cells. *Zentralbl. Bakterial. Mikrobiol. Hyg.*, B185(4-5): 427-429.

MODERN PLASTICS ENCYCLOPEDIA (1975) International Advertising Supplement 52: 10A, p. 697, New York McGraw-Hill Inc.

MONSANTO INDUSTRIAL CHEMICALS CO. (1979) Summary of the mutagenicity study, neurotoxicity study, teratology study, long-term feeding study and 90-day inhalation study which Monsanto has on the aryl phosphate, Saint-Louis, Missouri, Monsanto Industrial Chemicals Co. (Prepared for the US Environmental Protection Agency) (EPA-OTS document No. 40-7942057).

MUIR, D.C.G. (1984) Phosphate esters. In: Hutzinger, O., ed. *The handbook of environmental chemistry,* Berlin, Heidelberg, New York, Tokyo, Springer-Verlag, Vol. 3, Part C, pp. 41-66.

MUIR, D.C.G., GRIFT, N.P., & SOLOMON, J. (1980a) Determination of several triaryl phosphates in fish and sediment samples. *Can. Plains Proc.*, 9: 1-12.

MUIR, D.C.G., GRIFT, N.P., BLOUW, A.P., & LOCKHART, W.L. (1980b) Environmental dynamics of phosphate esters. I. Uptake and bioaccumulation of triphenyl phosphate by rainbow trout. *Chemosphere,* 9: 525-532.

MUIR, D.C.G., GRIFT, N.P., & SOLOMON, J. (1981) Extraction and cleanup of fish, sediment, and water for determination of triaryl phosphates by gas-liquid chromatography. *J. Assoc. Off. Anal. Chem.*, **64**: 79-84.

MUIR, D.C.G., GRIFT, N.P., & LOCKHART, W.L. (1982) Comparison of laboratory and field results for prediction of the environmental behavior of phosphate esters. *Environ. Toxicol. Chem.*, **1**: 113-119.

MUIR, D.C.G., YARECHEWSKI, A.L., & GRIFT, N.P. (1983a) Environmental dynamics of phosphate esters. III. Comparison of the bioconcentration of four triaryl phosphates by fish. *Chemosphere*, **12**: 155-166.

MUIR, D.C.G., TOWNSEND, B.E., & LOCKHART, W.L. (1983b) Bioavailability of six organic chemicals to *Chironous tentans* larvae in sediment and water. *Environ. Toxicol. Chem.*, **2**: 269-281.

NEELY, W.B., BRANSON, D.R., & BLAU, G.E. (1974) Partition coefficient to measure bioconcentration potential of organic chemicals in fish. *Environ. Sci. Technol.*, **8**: 1113-1115.

NEVINS, M.J. & JOHNSON, W.W. (1978) Acute toxicity of phosphate ester mixtures to invertebrates and fish. *Bull. environ. Contam. Toxicol.*, **19**: 250-256.

NOMEIR, A.A. & ABOU-DONIA, M.B. (1983) High-performanace liquid chromatographic analysis on radial compression column of the neurotoxic tri-o-cresyl phosphate and metabolites. *Anal. Biochem.*, **135**: 296-303.

OFSTAD, E.B. & SLETTEN, T. (1985) Composition and water solubility determination of a commercial tricresyl phosphate. *Sci. total Environ.*, **43**: 233-241.

PALAWSKI, D., BUCKLER, D.R., & MAYER, F.L. (1983) Survival and condition of Rainbow trout (*Salmo gairdneri*) after acute exposures to methyl parathion, triphenyl phosphate, and DEF. *Bull. environ. Contam. Toxicol.*, **30**: 614-620.

PEEREBOOM, J.W.C. (1960) The analysis of plasticizers by micro-adsorption chromatography. *J. Chromatogr.*, **4**: 323-328.

PEGUM, J.S. (1966) Contact dermatitis from plastics containing triaryl phosphtes. *Br. J. Dermatol.*, **78**: 626-631.

PICKARD, M.A., WHELIHAN, J.A., & WESTLAKE, D.W.S. (1975) Utilization of triaryl phosphates by a mixed bacterial population. *Can. J. Microbiol.*, **21**: 140-145.

PLAPP, F.W., Jr & TONG, H.H.C. (1966) Synergism of malathion and parathion against resistant insects: Phosphorus esters with synergistic properties. *J. econ. Entomol.*, **59**(1): 11-15.

RENBERG, L., SUNDSTROM, G., & SUNDH-NYGARD, K. (1980) Partition coefficients of organic chemicals derived from reversed phase thin layer chromatography. Evaluation of methods and application on phosphate esters, polychlorinated paraffins and some PCB-substitutes. *Chemosphere*, **9**: 683-691.

References

SAEGER, V.W., HICKS, O., KALEY, R.G., MICHAEL, P.R., MIEURE, J.P., & TUCKER, E.S. (1979) Environmental fate of selected phosphate esters. *Environ. Sci. Technol.*, 13: 840-844.

SASAKI, K., TAKEDA, M., & UCHIYAMA, M. (1981) Toxicity, absorption and elimination of phosphoric acid triesters by Killifish and Goldfish. *Bull. environ. Contam. Toxicol.*, 27: 775-782.

SASAKI, K., SUZUKI, T., TAKEDA, M., & UCHIYAMA, M. (1982) Bioconcentration and excretion of phosphoric acid triesters by Killifish *(Oryzeas laptipes)*. *Bull. environ. Contam. Toxicol.*, 28: 752-759.

SHELDON, L.S. & HITES, R.A. (1978) Organic compounds in the Delaware River. *Environ. Sci. Technol.*, 12: 1188-1194.

SHELDON, L.S. & HITES, R.A. (1979) Sources and movement of organic chemicals in the Delaware River. *Environ. Sci. Technol.*, 13: 574-579.

SITTHICHAIKASEM, S. (1978) Some toxicological effects of phosphate esters on rainbow trout and bluegill, Iowa State University (Ph. D. Thesis).

SMITH, M.I., EVOLVE, E., & FRAZIER, W.H. (1930) Pharmacological action of certain phenol esters with special reference to the etiology of so-called ginger paralysis. *Public Health Rep.*, 45: 2509-2524.

SMITH, M.I., ENGEL, E.W., & STOHLMAN, F.F. (1932) Further studies on the pharmacology of certain phenol esters with special reference to the relation of chemical constitution and physiologic action. *Natl. Inst. Health Bull.*, 160: 1-53.

SOBOTKA, T.J., BRODIE, R.E., ARNOLD, A., WEST, G.L., & O'DONNELL, M.W. (1986) Neuromotor function in rats during subchronic dietary exposure to triphenyl phosphate. *Neurobehav. Toxicol. Teratol.*, 8: 7-10.

SUGIYAMA, H. & TANAKA, K. (1982) [Investigation of trace organic chemicals in the sea water of the Tokyo Bay by gas chromatography mass-spectrometry.] *Bull. Kanagawa Prefect. environ. Center*, 4: 33-38 (in Japanese).

SUTTON, W.L., TERHAAR, C.J., MILLER, F.A., SCHERBERGER, R.F., RILEY, E.C., ROUDABUSH, R.L., & FASSETT, D.W. (1960) Studies on the industrial hygiene and toxicology of triphenyl phosphate. *Arch. environ. Health*, 1: 45-58.

SZYBALSKI, W. (1958) Special microbiological systems. II. Observations on chemical mutagenesis in microorganisms. *Ann. N.Y. Acad. Sci.*, 76: 475-489.

THEISS, J.C., STONER, G.D., SHIMKIN, M.B., & WEISBURGER, E.K. (1977) Test for carcinogenicity of organic contaminants of United States drinking waters by pulmonary tumor response in strain A mice. *Cancer Res.*, 37: 2717-2720.

TITTARELLI, P. & MASCHERPA, A. (1981) Liquid chromatography with graphite furnace atomic absorption spectrophotometric detector for speciation of organophosphorous compounds. *Anal. Chem.*, 53: 1466-1499.

US NIOSH (1980) *Industrial hygiene walk-through survey report on organophosphorus exposures at Rochester products division, General Motors Corporation, Rochester,* Cincinnati, Ohio, National Institute for Occupational Safety and Health (PB82-104530).

US NIOSH (1982) Industrial hygiene walk-through survey report on organophosphorus exposures at Chevron Chemical, Belle Chasse, Louisiana, Cincinnati, Ohio, National Institute for Occupational Safety and Health (Report No. IWA-89-10).

VAINIOTALO, S., VERKKALA, E., SAVOLAINEN, H., NICKELS, J., & ZITTING, A. (1987) Acute biological effects of commercial cresyl diphenyl phosphate in rats. *Toxicology,* 44(1): 31-44.

VASWANI, M., MAHAJAN, P.M., SETIA, R.C., & BHIDE, N.K. (1983) A simple colorimetric method for estimation of tricresyl phosphate in edible oils. *J. Oil Technol. Assoc. India,* 15(1): 12-13.

VEITH, G.D., DEFOE, D.L., & BERGSTEDT, B. (1979) Measuring and estimating the bioconcentration factor of chemicals in fish. *J. Fish Res. Board Can.,* 36: 1040-1048.

VICK, R.D., JUNK, G.A., AVERY, M.J., RICHARD, J., & SVEC, H.J. (1978) Organic emissions from combustion of combination coal/refuse to produce electricity. *Chemosphere,* 7: 893-902.

WAGEMANN, R., GRAHAM, B., & LOCKHART, W.L. (1974) *Studies on chemical degradation and fish toxicity of a synthetic triaryl phosphate lubrication oil, IMOL S-140,* Ottawa, Environment Canada, Fisheries and Marine Service (Technical Report No. 486).

WAKABAYASHI, A. (1980) [Environmental pollution and toxicological aspects of a triaryl phosphate synthetic oil]. *Annual Report of the Tokyo Metropolitan Research Institute for Environmental Protection,* 11: 110-113 (in Japanese).

WELSH, J.J., COLLINS, T.F.X., WHITBY, K.E., BLACK, T.N., & ARNOLD, A. (1987) Teratogenic potential of triphenyl phosphate in Sprague-Dawley (Spartan) rats. *Toxicol. ind. Health,* 3(3): 357-369.

WHO (1990) Environmental Health Criteria 110: Tricresyl phosphate, Geneva, World Health Organization.

WILLIAMS, D.T. & LEBEL, G.L. (1981) A national survey of tri(haloalkyl)-, trialkyl-, and triaryl phosphates in Canadian drinking water. *Bull. environ. Contam. Toxicol.,* 27: 450-457.

WILLIAMS, D.T., NESTMANN, E.R., LEBEL, G.L., BENOIT, F.M., & OTSON, R. (1982) Determination of mutagenic potential and organic contaminants of Great Lakes drinking water. *Chemosphere,* 11: 263-276.

WILLS, J.H., BARRON, K., GROBLEWSKI, G.E., BENITZ, K.F., & JOHNSON, M.K. (1979) Does triphenyl phosphate produce delayed neurotoxic effects? *Toxicol. Lett.,* 4: 21-24.

References

WINDHOLZ, M., ed. (1983) *The Merck Index*, 10th ed., Rahway, New Jersey, Merck and Co., Inc.

WOLFE, N.L. (1980) Organophosphate and organophosphorothionate esters: Application of linear free energy relationships to estimate hydrolysis rate constants for use in environmental fate assessment. *Chemosphere*, 9: 571-579.

WONG, P.T.S. & CHAU, Y.K. (1984) Structure-toxicity of triaryl phosphates in freshwater algae. *Sci. total Environ.*, 32(2): 157-65.

YASUDA, H. (1980) [Concentration of organic phosphorus pesticides in the atmosphere above the Dogo plain and Ozu basin.] *J. Chem. Soc. Jpn*, 1980: 645-653 (in Japanese).

ZEIGER, E., ANDERSON, B., HAWORTH, S., LAWLOR, T., MORTELMANS, K., & SPECK, W. (1987) *Salmonella* mutagenicity tests: III. Results from the testing of 255 chemicals. *Environ. Mutagen.*, 9(Suppl. 9): 1-110.

ZITKO, V. (1980) Proceedings of the 6th Annual Aquatic Toxicity Workshop, *Can. Tech. Rep. Fish. aquat. Sci.*, 575: 234-265.

RESUME

1. Identité, propriétés physiques et chimiques, méthodes d'analyse

Le phosphate de triphényle (TPP) est une substance cristalline, ininflammable, inexplosible et incolore. Son coefficient de partage entre l'octanol et l'eau (log de P_{ow}) est de 4,61-4,76. A la température ambiante ordinaire, il s'hydrolyse rapidement en milieu alcalin pour donner du phosphate de diphényle et du phénol, mais l'hydrolyse est très lente en milieu acide ou neutre.

Pour l'analyse, la méthode choix est la chromatographie gaz-liquide, avec détection au moyen d'un dispositif sensible à l'azote/phosphore ou par photométrie de flamme. La limite de détection dans l'eau est d'environ 20 ng/litre.

2. Sources d'exposition humaine et environnementale

Le phosphate de triphényle est produit à partir de l'oxychlorure de phosphore et du phénol. Il est utilisé comme retardateur de flamme dans les résines phénoliques et les résines à base d'oxydes de phénylène que l'on utilise pour la fabrication de pièces d'automobiles et de l'appareillage électrique; on l'emploie également comme plastifiant ininflammable dans l'acétate de cellulose servant à la confection des pellicules photographiques. Il entre également dans la compositision des liquides hydrauliques et des huiles lubrifiantes à côté d'un certain nombre d'autres usages de moindre importance.

On peut considérer qu'en utilisation normale, la population dans son ensemble n'encourt qu'une exposition minime.

3. Transport, distribution et transformation dans l'environnement

Les phosphates de triaryle pénètrent dans le milieu aquatique par suite de fuites de liquides hydrauliques, de la lixiviation de certains plastiques et en faibles

Résumé

quantités, lors des divers processus de fabrication. En raison de sa faible solubilité dans l'eau et de son coefficient d'adsorption au sol relativement élevé, le phosphate de triphényle se fixe rapidement sur les sédiments de rivières ou des étangs. En milieu aquatique, il subit une biodégradation rapide.

La dégradation du phosphate de triphényle comporte une hydrolyse enzymatique par étapes en orthophosphate et phénol.

Les facteurs de bioconcentration mesurés chez plusieurs espèces de poissons vont de 6 à 18 900 et la demi-vie d'élimination varie de 1,2 à 49,6 heures.

La libération de cette substance dans l'air des unités de production constitue une source d'exposition humaine sur les lieux de travail. La combustion des matières plastiques et la volatilisation du phosphate de triphényle à partir de ces substances ou de la surface de l'eau peut également constituer une voie importante de pénétration dans l'atmosphère.

4. Niveaux dans l'environnement et exposition humaine

On trouve un peu partout du phosphate de triphényle dans l'air, l'eau, les sédiments et les organismes aquatiques, mais les prélèvements effectués n'en contiennent que de faibles quantités. Les teneurs les plus fortes qui aient été signalées sont de 23,2 ng/m^3 dans l'air, 7900 ng/litre dans des cours d'eau, 4000 ng/g dans des sédiments et de 600 ng/g dans le poisson.

5. Effets sur les êtres vivants dans leur milieu naturel

La croissance des algues est complètement inhibée à des concentrations de 1 mg/litre ou davantage mais elle est en revanche stimulée à des concentrations plus faibles (0,1 et 0,05 mg/litre). L'activité de la nitrogénase d'*Anabaena flos-aquae* diminue à mesure que la dose augmente, passant de 84% à 0,1 mg/litre à 68% à 5,0 mg/litre.

Le phosphate de triphényle est, parmi les phosphates de triaryle, celui qui présente la plus forte toxicité aiguë vis-à-vis des poissons, des crevettes et des daphnies. L'indice de toxicité aiguë de ce phosphate pour le poisson (CL_{50} à 96 h) va de 290 mg/litre pour *Lepomis macrochirus,* à 0,36 mg/litre pour la truite arc-en-ciel. La grande différence entre la truite et les vairons du genre Pimephales en ce qui concerne les valeurs de la CL_{50}, pourraient être due à des différences dans leur aptitude à métaboliser le phosphate de triphényle. Parmi les effets sublétaux observés chez les poissons, on peut citer des anomalies morphologiques telles que congestion, dégénérescence et hémorragie au niveau des petits vaisseaux sanguins (essentiellement branchiaux) ainsi que des anomalies de comportement. L'immobilité des poissons exposés à 0,21-0,29 mg/litre de phosphate de triphényle a complètement disparu dans les sept jours qui ont suivi le changement d'eau.

6. Effets sur les animaux d'expérience et les systèmes d'épreuves *in vitro*

On estime que la DL_{50} par voie orale est supérieure à 6,4 g/kg chez le rat et à 2,0 g/kg chez le poulet.

Des doses de phosphate de triphényle allant de 0,5 à 2 g/kg ont été bien tolérées par des lapins après injection intramusculaire et par des poulets après administration orale. Lors d'une étude d'alimentation de 35 jours, on a observé après administration de cette substance à des rats Holtzman mâles, une réduction du gain de poids corporel et une augmentation du poids du foie.

On n'a pas observé d'effets tératogènes chez des rats Sprague-Dawley à des doses allant jusqu'à 690 mg/kg de poids corporel. On n'a pas publié d'études concernant la reproduction.

On ne dispose pas de données sur la mutagénicité du phosphate de triphényle qui résultent d'épreuves correctement validées et il n'y a pas eu non plus d'études de cancérogénicité convenables.

Après des injections sous-cutanées de phosphate de triphényle à des chats (jusqu'à 1 g/kg) on n'a pas observé de neurotoxicité retardée; on n'en a pas observé non plus après une étude de 4 mois sur des rats Sprague-Dawley qui en recevaient dans leur alimentation des doses allant jusqu'à 1% de la ration.

Résumé

Aucun effet immunotoxique n'a été signalé après une étude de 120 jours pendant laquelle des rats ont reçu du phosphate de triphényle dans leur nourriture à des doses allant jusqu'à 1%.

7. Effets sur l'homme

On a signalé une réduction statistiquement significative de la cholinestérase érythrocytaire chez certains travailleurs, mais aucun signe d'affection neurologique n'a été relevé chez des ouvriers qui travaillaient dans une unité de production de phosphate de triphényle. On n'a pas signalé non plus de neurotoxicité retardée parmi les cas d'intoxication par le phosphate de triphényle. On a décrit des cas de dermatite de contact dus au phosphate de triphényle.

EVALUATION DES RISQUES POUR LA SANTE HUMAINE ET DES EFFETS SUR L'ENVIRONNEMENT

1. Evaluation des risques pour la santé humaine

Les données tirées des études sur l'animal montrent que le phosphate de triphényle est peu toxique. Appliqué sur la peau d'animaux de laboratoire, il ne produit pas d'irritation. On estime que le phosphate de triphényle n'est pas neurotoxique pour l'homme ni l'animal, bien qu'un premier rapport ait pu affirmer le contraire. Lors d'une étude de 90 jours sur des rats, on a évalué à 690 mg/kg par jour la dose sans effet nocif observable pour les mères et leur descendance. L'exposition professionnelle et l'exposition de la population dans son ensemble demeurent à un faible niveau.

Le phosphate de triphényle n'est pas mutagène.

Selon des données disponibles, il ne présente aucun danger pour l'homme.

1.1 Niveaux d'exposition

Il y a probablement un risque d'exposition de la population générale au phosphate de triphényle par l'intermédiaire des divers compartiments de l'environnement et notamment par l'eau de consommation. Toutefois les concentrations de phosphate de triphényle mesurées dans de l'eau de boisson au Canada et aux Etats-Unis se sont révélées extrêmement faibles. On en a souvent décelé la présence dans l'air des villes mais à faibles concentrations. On a pu craindre que l'échauffement des sièges d'automobiles en vinyle, lorsque la température extérieure est très élevée, puisse conduire à la vaporisation du phosphate de triphényle utilisé comme plastifiant, mais on ne dispose d'aucune donnée sur les concentrations présentes à l'intérieur des véhicules. Lors d'une enquête portant sur la teneur des tissus adipeux humains en phosphates de triaryle, on n'a pas décelé la présence de phosphate de triphényle. On ne dispose pas de données suffisantes pour se faire une idée de l'importance de l'exposition de la population générale au phosphate de triphényle. On a signalé des concentrations importantes de ce produit dans l'air d'une unité de production (0,5 à 29,6 mg/m^3) mais on ne dispose pas de chiffres récents. Il serait bon d'avoir davantage de données sur l'exposition professionnelle au phosphate de triphényle dans les unités de production.

1.2 Effets toxiques

Le profil de toxicité du phosphate de triphényle ne permet guère d'évaluer de façon complète le danger qu'il représente.

On a noté aucun signe d'activité mutagène chez les bactéries ni d'ailleurs d'activité cancérogène, en se basant pour cela sur une étude relative à une seule espèce animale. L'expérimentation animale n'a pas pu jusqu'ici, mettre en évidence une neurotoxicité retardée attribuable à cette substance. Lors d'une étude d'alimentation de 35 jours sur des rats, on a relevé à la dose de 5 g/kg, une réduction du gain de poids corporel et une augmentation du poids du foie. On ne possède pas de données suffisantes sur les effets qu'il pourrait exercer sur la fonction de reproduction (gonades, fécondité, parturitions, croissance et développement de la progéniture).

On a décrit des cas de dermatite de contact attribuable au phosphate de triphényle.

2. Evaluation des effets sur l'environnement

Dans l'eau, la concentration du phosphate de triphényle est faible et il est peu probable qu'il exerce des effets toxiques sur les organismes aquatiques. Il peut y avoir mortalité locale par suite du déversement accidentel de liquides hydrauliques contenant du phosphate de triphényle. Cependant, comme ce phosphate s'élimine rapidement des tissus pisciaires lorsque cesse l'exposition et que les facteurs de bioconcentration sont moyens, on ne pense pas qu'il y ait véritablement risque de bioaccumulation.

On a fait état de fortes concentrations de phosphate de triphényle dans les sédiments proches des unités de production. Il a été montré en outre que le phosphate de triphényle lié aux sédiments pouvait être fixé par un organisme qui y était présent mais on ne possède aucune donnée de toxicité sur les espèces qui vivent dans les sédiments ou qui s'en nourrissent. Reste que des effets sur les populations aquatiques sont possibles.

2.1 Niveaux d'exposition

Dans les régions très industrialisées, les prélèvements effectués dans l'air, dans les eaux superficielles, dans le sol, les sédiments et parmi les organismes aquatiques indiquent la présence de phosphate de triphényle. La concentration la plus élevée qui ait été signalée dans des effluents industriels était de 16 µg/litre; dans un cours d'eau, elle était de 7,9 µg/litre. Si l'on prend en considération la biodégradation rapide du phosphate de triphényle dans le milieu aquatique, il est peu probable que les concentrations que l'on rencontre normalement puissent se révéler nocives pour les organismes qui y vivent. Toutefois la décharge dans des mares de déchets de garnitures de sièges en vinyle traité par du phosphate de triphényle, pourrait donner lieu à des concentrations mortelles pour les poissons.

2.2 Effets toxiques

Parmi les divers phosphates de triaryle, le phosphate de triphényle est celui dont la toxicité aiguë est la plus forte pour les poissons, les crevettes et les daphnies. La CL_{50} à 96 h varie de 0,36 mg/litre pour la truite arc-en-ciel à 290 mg/litre pour *Lepomis macrochirus*. Les salmonidés sont en général sensibles au phosphate de triphényle mais on a constaté qui ni la croissance ni la survie des alevins de truite arc-en-ciel ne souffraient d'une exposition à cette substance à la concentration de 0,0014 mg/litre. Chez des poissons rouges exposés à du phosphate de triphényle, on a constaté un certain nombre d'anomalies histologiques: congestion, dégénérescence et hémorragie au niveau des petits vaisseaux sanguins, principalement les veinules et les capillaires. Cette pathologie vasculaire est plus prononcée au niveau des branchies.

La présence de concentrations de phosphate de triphényle de l'ordre de 1 mg/litre ou davantage a complètement

Evaluation

inhibé la croissance de certaines algues alors que des concentrations plus faibles (0,1 et 0,05 mg/litre) avaient l'effet contraire. L'activité de la nitrogénase d'*Anabaena flos-aquae* a été sensiblement réduite, même à la concentration de 0,1 mg/litre.

RECOMMANDATIONS

1. Recommandations relatives aux recherches à effectuer

    a) Etudes à entreprendre sur la sensibilisation cutanée.

    b) Nécessité d'une étude de cytogénicité *in vitro*.

    c) Nécessité d'études pharmacocinétiques selon les différentes voies d'absorption.

RESUMEN

1. **Identidad, propiedades físicas y químicas y métodos analíticos**

    El trifenilfosfato (TFF) es una sustancia no inflamable, no explosiva, incolora y cristalina. Su coeficiente de reparto en octanol y agua (log P_{oa}) es de 4,61-4,76. A temperatura ambiente normal se hidroliza rápidamente en solución alcalina, dando difenilfosfato y fenol, y muy lentamente en soluciones ácidas o neutras.

    El método analítico más apropiado es la cromatografía gas-líquido con un detector sensible al nitrógeno-fósforo o uno fotométrico de llama. El límite de detección en el agua es de unos 20 ng/litro.

2. **Fuentes de exposición humana y ambiental**

    El TFF se fabrica a partir de oxicloruro de fósforo y fenol. Se utiliza como pirorretardante en resinas fenólicas y de óxido de fenileno en la producción de componentes eléctricos y del automóvil y como plastificante no inflamable en acetato de celulosa para películas fotográficas. También es un componente de fluidos hidráulicos o aceites lubricantes y tiene otros usos de menor importancia.

    La exposición de la población general por el uso normal puede considerarse mínima.

3. **Transporte, distribución y transformación en el medio ambiente**

    Los triarilfosfatos entran en el medio acuático principalmente por escapes de fluidos hidráulicos, así como por lixiviación a partir de los plásticos y, en menor medida, a partir de los procesos de fabricación. A causa de su baja solubilidad en agua y su coeficiente de adsorción en el suelo relativamente alto, el TFF se adsorbe con rapidez en los sedimentos de los ríos (o de las charcas). Su biodegradación en el medio acuoso es rápida.

    El TFF se degrada mediante una hidrólisis enzimática escalonada que lo divide en ortofosfato y componentes fenólicos.

Los factores de bioconcentración (FBC) medidos en varias especies de peces oscilan entre 6 y 18 900, y la semivida de depuración va de 1,2 a 49,6 h.

La liberación de TFF desde los lugares de producción al aire representa una fuente de exposición humana en el ambiente de trabajo. La combustión de plásticos y la volatilización a partir de ellos o de las superficies acuáticas también pueden ser importantes vías de ingreso en la atmósfera.

4. Niveles medioambientales y exposición humana

El TFF se ha detectado con frecuencia en el aire, el agua, los sedimentos y los organismos acuáticos, pero los niveles en muestras medioambientales son bajos. Los niveles máximos detectados son de 23,2 ng/m^3 en el aire, 7900 ng/litro en agua de río, 4000 ng/g en sedimentos y 600 ng/g en peces.

5. Efectos sobre los seres vivos del medio ambiente

Las concentraciones de TFF iguales o superiores a 1 mg/litro inhiben completamente el crecimiento de las algas, pero la concentraciones más bajas (0,1 y 0,05 mg/litro) lo estimulan. La actividad de la nitrogenasa de *Anabaena flos-aquae* decrece en función de la dosis desde 84% a 0,1 mg/litro hasta 68% a 5,0 mg/litro.

De los distintos triarilfosfatos, el TFF es el más tóxico para los peces, los camarones y los dáfnidos. El índice de toxicidad aguda del TFF para los peces (CL$_{50}$ 96 h) oscila entre 290 mg/litro en *Lepomis macrochirus* y 0,36 mg/litro en la trucha arco iris. La gran diferencia entre los valores de CE$_0$ de la trucha y de *Pimephales promelas* puede deberse a su distinta capacidad para metabolizar el TFF. Entre los efectos subletales que produce en los peces figuran anomalías morfológicas como congestión, degeneración y hemorragias de los vasos sanguíneos más pequeños (principalmente de las branquias) y anomalías en el comportamiento. La inmovilidad de los peces expuestos a concentraciones de 0,21-0,29 mg/litro desapareció totalmente al cabo de siete días cuando se los puso en agua limpia.

Resumen

6. Efectos en los animales de experimentación y en sistemas de prueba *in vitro*

Se ha calculado que por vía oral la DL_{50} del TFF en ratas es > 6,4 g/kg y en pollos es > 2,0 g/kg.

Los conejos y los pollos toleraron bien dosis de TFF de 0,5 a 2,0 g/kg por vía intramuscular y oral respectivamente. En un estudio de alimentación de 35 días en machos de rata Holtzman, con una dosis de TFF se observó una disminución en la ganancia de peso corporal y un aumento del peso del hígado.

Con dosis de TFF de hasta 690 mg/kg de peso corporal no se produjeron efectos teratogénicos en ratas Sprague-Dawley. No se conocen estudios sobre la reproducción.

No hay datos acerca de la capacidad mutagénica del TFF obtenidos en ensayos bien contrastados y no se han hecho estudios adecuados de carcinogenicidad.

La aplicación a gatos de una única dosis subcutánea de TFF (de hasta 1 g/kg) no causó neurotoxicidad diferida; tampoco se observó en un estudio de cuatro meses en ratas Sprague-Dawley con dosis de hasta el 1% en el alimento.

No se comunicaron efectos inmunotóxicos tras un estudio de 120 días en ratas a las que se administraron dosis de hasta el 1% en el alimento.

7. Efectos en la especie humana

Se ha comunicado que, si bien algunos trabajadores de instalaciones de producción de TFF han mostrado una reducción estadísticamente significativa de la colinesterasa eritrocítica, no hay manifestaciones de enfermedades neurológicas. No hay informes de neurotoxicidad diferida en casos de intoxicación por TFF. Se han descrito casos de dermatitis de contacto debida al TFF.

EVALUACION DE LOS RIESGOS PARA LA SALUD HUMANA Y DE LOS EFECTOS EN EL MEDIO AMBIENTE

1. Evaluación de los riesgos para la salud humana

   Los datos obtenidos en animales indican que el TFF tiene una toxicidad baja. No produce efectos irritantes en la piel de los animales. A pesar de un primer informe en sentido contrario, no se considera que el TFF sea neurotóxico para los animales o el hombre. El nivel sin efecto adverso observado fue, en las madres y en las crías, de 690 mg/kg al día en un estudio de 90 días realizado en ratas. La exposición al TFF es baja tanto en la población general como en los trabajadores.

   El TFF no tiene efectos mutagénicos.

   Los datos disponibles indican que no entraña peligro para los seres humanos.

1.1 Niveles de exposición

   Puede considerarse probable la exposición de la población general al TFF por conducto de distintos medios ambientales, incluida el agua de bebida. Se han medido concentraciones extraordinariamente bajas de TFF en el agua de bebida del Canadá y los EE.UU. Con frecuencia se ha detectado TFF en el aire urbano, aunque los niveles son bajos. Se ha hablado de vaporización de TFF al calentarse el vinilo de la tapicería de los automóviles, pero no se dispone de datos sobre la concentración en los coches. En un estudio de los triarilfosfatos en el tejido adiposo humano no se detectó TFF. Estos datos no bastan para evaluar la importancia de la exposición de la población general al TFF.

   Se ha comunicado la existencia de concentraciones importantes (0,5-29,6 mg/m^3) en el aire de unas instalaciones de producción de TFF, pero no se dispone de cifras recientes. Se necesitan más datos sobre la exposición profesional al TFF en los lugares de fabricación.

Evaluación

1.2 Efectos tóxicos

Los datos de que se dispone sobre la toxicidad del TFF son totalmente insuficientes para una valoración completa del riesgo que representa.

No hay pruebas de que el TFF tenga actividad mutagénica en bacterias ni de que tenga actividad carcinogénica, de acuerdo con un estudio sobre una especie animal. De momento no se han obtenido pruebas de que el TFF cause neurotoxicidad diferida en animales de experimentación. En un estudio de alimentación de 35 días en ratas se observó una disminución en la ganancia de peso corporal y un aumento de peso del hígado con una dosis de 5 g/kg. No se dispone de datos adecuados en cuanto a los efectos del TFF en la reproducción, es decir, en la función de las gónadas, la fertilidad, el parto y el crecimiento y desarrollo de la descendencia.

Se han descrito casos de dermatitis de contacto causada por el TFF.

2. Evaluación de los efectos en el medio ambiente

Las concentraciones de TFF en el agua ambiental son bajas y los efectos tóxicos en los organismos acuáticos son poco probables. Los vertidos de fluidos hidráulicos con TFF podrían tener efectos letales a nivel local. Puesto que el TFF se elimina rápidamente de los tejidos de los peces al terminar la exposición y los factores de bioconcentración son moderados, no se considera que la bioacumulación sea un peligro.

Se ha informado de la presencia de altas concentraciones de TFF en sedimentos cercanos a instalaciones de producción. Se ha demostrado que cierto organismo que vive en los sedimentos puede utilizar el TFF fijado por éstos, pero no hay datos de toxicidad sobre las especies que viven en los sedimentos o se alimentan de ellos. Existe, por consiguiente, la posibilidad de efectos en las comunidades acuáticas.

2.1 Niveles de exposición

El TFF se halla en el aire, el agua superficial, el suelo, los sedimentos y los organismos acuáticos recogidos en zonas muy industrializadas. La concentración más alta de TFF en efluentes de aguas industriales que se ha comunicado es de 16 µg/litro, mientras que en aguas fluviales es de 7,9 µg/litro. Teniendo en cuenta la rápida biodegradación del TFF en el medio acuoso, es poco probable que las concentraciones normales de TFF en él afecten de manera adversa a los organismos acuáticos. Sin embargo, si se arrojase a una charca tejido de tapicería de vinilo tratado con TFF se produciría una concentración suficientemente alta para matar a los peces.

2.2 Efectos tóxicos

Entre los diferentes triarilfosfatos, el TFF es el compuesto más tóxico para peces, camarones y dáfnidos. Los valores de la CL_{50} en 96 horas de TFF para los peces varían entre 0,36 mg/litro en la trucha arco iris y 290 mg/litro en *Lepomis macrochirus*. Aunque los salmónidos en general son sensibles al TFF, el crecimiento y la supervivencia de los alevines de trucha arco iris no se vieron afectados cuando éstos se expusieron a una concentración de TFF de 0,0014 mg/litro. Los ejemplares de *Carassius auratus* expuestos al TFF presentaron lesiones histopatológicas consistentes en congestión, degeneración y hemorragia de los vasos sanguíneos pequeños, principalmente vénulas y capilares. Esta patología vascular es más pronunciada en las branquias.

Las concentraciones de TFF iguales o superiores a 1 mg/litro inhibieron completamente el crecimiento de las algas, pero a concentraciones más bajas (0,1 y 0,05 mg/litro) lo estimularonn. La actividad de la nitrogenasa de *Anabaena flos-aquae* se redujo de forma significativa, incluso a una concentración de 0,1 mg/litro.

RECOMENDACIONES

1. **Recomendaciones para futuras investigaciones**

    a) Se deberían realizar estudios de sensibilización cutánea.

    b) Es necesario realizar un estudio de citogenicidad *in vitro*.

    c) Se precisan estudios farmacocinéticos de las diferentes vías.

www.ingramcontent.com/pod-product-compliance
Ingram Content Group UK Ltd.
Pitfield, Milton Keynes, MK11 3LW, UK
UKHW021307180426
11947UKWH00015B/1064